The Theory
of Horticulture

JOHN LINDLEY

CAMBRIDGE
UNIVERSITY PRESS

CAMBRIDGE UNIVERSITY PRESS

Cambridge, New York, Melbourne, Madrid, Cape Town,
Singapore, São Paolo, Delhi, Tokyo, Mexico City

Published in the United States of America by Cambridge University Press, New York

www.cambridge.org
Information on this title: www.cambridge.org/9781108037242

This edition first published 1840
This digitally printed version 2011

ISBN 978-1-108-03724-2 Paperback

THE

THEORY

OF

HORTICULTURE;

OR,

AN ATTEMPT TO EXPLAIN

THE PRINCIPAL OPERATIONS OF GARDENING

UPON

PHYSIOLOGICAL PRINCIPLES.

BY

JOHN LINDLEY, Ph.D. F.R.S.

VICE-SECRETARY OF THE HORTICULTURAL SOCIETY OF LONDON,
AND PROFESSOR OF BOTANY IN UNIVERSITY COLLEGE.

" Though I am very sensible that it is from long experience chiefly that we are to
expect the most certain rules of practice, yet it is withal to be remembered that
the likeliest method to enable us to make the most judicious observations, and to
put us upon the most probable means of improving any art, is to get the best in-
sight we can into the nature and properties of those things which we are desirous
to cultivate and improve." — *Hales's Vegetable Statistics*, i. 376.

LONDON:

PRINTED FOR

LONGMAN, ORME, BROWN, GREEN, AND LONGMANS
PATERNOSTER-ROW.

1840.

London:
Printed by A. Spottiswoode,
New-Street-Square.

TO

THE MEMORY

OF

THOMAS ANDREW KNIGHT.

A 2

PREFACE.

THIS book is written in the hope of providing the intelligent gardener, and the scientific amateur, correctly, with the rationalia of the more important operations of Horticulture; in the full persuasion that, if the physiological principles on which such operations, of necessity, depend, were correctly appreciated by the great mass of active-minded persons now engaged in gardening in this country, the grounds of their practice would be settled upon a more satisfactory foundation than can at present be said to exist. It is, I confess, surprising to me, that the real nature of the vital actions of plants, and of the external forces by which they are regulated, should be so frequently misapprehended even among writers upon Horticulture; and that ideas relating to such matters, so very incorrect as we frequently find them to be, should obtain among intelligent men, in the present state of what I may be permitted to call horticultural physiology.

There must be a great want of sound knowledge
of this subject, when we find an author, who has
made himself distinguished in the history of English gardening, giving it as his opinion, "that the
weak drawn state of forced Asparagus in London
is occasioned by the action of the dung immediately upon its roots!"

It does not seem possible to account for this
in any other way than by referring it to the
want of some short guide to the horticultural application of vegetable physiology, unmixed with
other things; and so arranged that the intimate
connexion of one branch of practice with another, and of the whole with a few well ascertained facts upon which every thing else depends,
may be distinctly perceived from a single point
of view.   The admirable papers of Mr. Knight
are scattered through the *Horticultural Transactions ;* and the writings of other physiologists are
dispersed through so many different works, that the
labour of finding them, when wanted, is greater than
is willingly undertaken even by those who have
access to ample libraries.   With regard to general
works on Horticulture, it is very far from my wish
to say one word in disparagement of the many excellent publications upon this subject which have

already appeared in this country ; on the contrary, the improved state of gardening among us may be reasonably ascribed to the influence of some of these valuable works : but it must be admitted that the true principles of physiology are not, in such books, so separated from the details of routine on the one hand, or so applied to them on the other, as to be readily understood by those who want either the skill or the inclination to distinguish empirical directions from rules which are plainly founded upon the very nature of things. I must also be permitted to observe that, although results are correctly stated in such books, they are not unfrequently referred to wrong causes.

In preparing the following pages for the press, my anxious desire has been to strike out all unnecessary matter, even although it may be required to complete the physiological explanation of common facts ; and to introduce little beyond that which every gardener can verify for himself. Vegetable anatomy is no doubt the foundation of all correct views of physiological action ; chemistry is of the first importance, when the general functions of plants are considered in a large and general way ; and electricity probably exercises an important influence over the vital actions of all living things.

But these are the refinements of science, belonging to the philosopher in his laboratory, and not to the worker in gardens ; they are indispensable to the correct appreciation of physiological phenomena, but not to the application of those phenomena to the arts of life ; electricity, in particular, appears to me, in the present imperfect state of our knowledge of its relation to vegetable functions, altogether incapable of forming a part of any horticultural theory.

What the gardener wants is, not a treatise upon botany, nor a series of speculations upon the possible nature of the influence on plants of all existing forces, nor an elaborate account of chemical agencies inappreciable by his senses and obscurely indicated by their visible results; but an intelligible explanation, founded upon well ascertained facts which he can judge of by his own means of observation, of the general nature of vegetable actions, and of the causes which, while they control the powers of life in plants, are themselves capable of being regulated by himself. The possession of such knowledge will necessarily teach him how to improve his methods of cultivation, and lead him to the discovery of new and better modes.

It is very true that ends of this kind are often

brought about by accident, without the smallest design on the part of the gardener ; and there are, doubtless, many men of uncultivated or idle minds, who think waiting upon Providence much better than any attempt to improve their condition by the exertion of their reasoning faculties. For such persons books are not written.

I hope that what has now been said will not lead any one to suppose that this sketch is offered to the reader as a complete theory of Horticulture in all its varied branches ; such a work would be alike tedious to the author and the reader, and, I fear, as unprofitable ; for, if a gardener, when once made acquainted with the general principles of science, has not the skill to apply them to each particular case, it is to be feared that no disquisition, however elaborate, would enable him to do so. So far has it been from my intention to enter into subordinate details, that I have carefully avoided them, from a fear of complicating the subject, and making that obscure which in itself is sufficiently clear. All that a physiologist has really to do with Horticulture is, to explain the general nature of the vital actions of a plant, and the manner in which these are commonly applied to the arts of cultivation ; if he quits this ground, he extends his limits so much that there is no longer a horizon in view. No one, indeed, could advantageously investigate the minor points of cultivation in all their branches, unless he were both a good physiologist and a practical gardener of the greatest experience, a combination of qualifications which no man has ever yet possessed, and to which I, most assuredly, have not the shadow of pretension.

In conclusion, let me, in impressing upon the minds of gardeners the importance of attending to first principles, also caution them against attempting to apply them, except in a limited manner, and by way of safe experiment, until they fully understand them. The difference between failure and success, in practice, usually depends upon slight circumstances, very easily overlooked, and not to be anticipated beforehand, even by the most skilful ; their importance is often unsuspected till an experiment has failed, and may not be discovered till after many unsuccessful attempts, during which more mischief may be done by extensive failures than the result is worth when attained. No man understood this better than the late Mr. Knight, the best horticultural physiologist that the world has seen, whose experiments

were conducted with a skill and knowledge which
few can hope to equal. So fully was he aware of
the uncertain issue of experimental investigations
in Horticulture, that he thought it necessary, in
recommending a new mode of cultivating the Pine-
apple, and in objecting to methods at that time
commonly in use, to express himself in the following words : — "I beg it to be understood that I
condemn the machinery only which our gardeners
employ, and that I admit most fully their skill in
the application of that machinery to be very superior to that which I myself possess. Nor do I
mean, in the slightest degree, to censure them for
not having invented better machinery, for it is
their duty to put in practice that which they have
learned ; and, having to expend the capital of
others, they ought to be cautious in trying expensive experiments, of which the results must necessarily be uncertain ; and, I believe, a very able and
experienced gardener, after having been the in-
ventor of the most perfect machinery, might, in
very many instances, have lost both his character
and his place before he had made himself sufficiently acquainted with it, and consequently become
able to regulate its powers."

ERRATUM.

Page 17. line 20. *for the paragraph beginning* " But in others," &c., *read* " But in others the circle occupied by these organs must be very much greater than that of the branches."

CONTENTS.

BOOK II.

THE

THEORY

OF

HORTICULTURE.

1. Horticulture is that branch of knowledge which relates to the cultivation, multiplication, and amelioration of the Vegetable Kingdom. It divides into two branches, which, although mutually dependent, are, in fact, essentially distinct : the art and the science. Under the art of horticulture is comprehended whatever concerns the mere manner of executing the operations connected with cultivation, multiplication, and amelioration ; the science explains the reasons upon which practice is founded. It is to the consideration of the latter subject that the following pages are dedicated.

2. It must have been remarked by all intelligent observers, that in the majority of works upon horticultural subjects, the numerous directions given in any particular ramification into which the art is susceptible of being divided, are held together by no bond of union, and that there is no explanation

B

2 THE THEORY OF HORTICULTURE.

of their connexion with general principles, by which alone the soundness of this or that rule of practice may be tested; the reader is therefore usually obliged to take the excellence of one mode of cultivation and the badness of another, upon the good faith of gardening authors, without being put into possession of any laws by which they may be judged of beforehand. Horticulture is by these means rendered a very complicated subject, so that none but practised gardeners can hope to pursue it successfully; and, like all empirical things, it is degraded into a code of peremptory precepts.

3. It will nevertheless be found, if the subject is carefully investigated, that in reality the explanations of horticultural operations are simple, and free from obscurity; provided they are not encumbered with speculations, which, however interesting they may be in theory, are only perplexing in practice, in the present state of knowledge. When, for example, chemical illustrations, unless of the simplest kind, or minute anatomical questions, or references to the agency of the electrical fluid, are discussed, the subject becomes embarrassed with considerations which are too refined for the apprehension of the majority of readers of gardening works, and which have little obvious application to practical purposes. Instead, therefore, of introducing points of obscure or doubtful application, or such as are not absolutely requi-

site for the explanation of phenomena, all which necessarily tend to complicate the theory of horticulture, it seems better strictly to confine our attention to the action of the simplest vital forces; for the general nature of these has been undoubtedly ascertained, and is easily understood by every class of readers. It is certain, for instance, that plants breathe, digest, and perspire; but it may be a question whether the exact nature of their respiration, digestion, and perspiration is beyond all further explanation; it is therefore better to limit our consideration to the naked fact, which is all that it imports the gardener to know, without inquiring too curiously into those phenomena. For it must always be remembered that the object of a work like the present is not to elucidate the laws of vegetable life in all their obscure details, but to teach, to those acquainted with the art of gardening, what the principles are upon which their practice is founded.

4. In order to attain this end it is necessary, in the first place, to explain briefly, but distinctly, the nature of those vital actions which have a direct reference to cultivation; omitting every thing that tends to embarrass the subject or which is not susceptible of a direct practical application; and in the next place, to show how those facts bear upon the routine of practice of the horticulturist, by making them explain the reason of the treat-

ment which is employed in various branches of the
gardener's art.

5. The first part of this work will therefore embrace the principal laws and facts in vegetable
physiology, as deduced from the investigations of the
botanist ; and the second the application of those
laws to practice, as explained by the experience of
the horticulturist.

6. If the laws comprehended in the first book
are correctly explained, and the facts connected
with them rightly interpreted, they must necessarily afford, in all cases, the reasons why one kind
of cultivation is better than another ; and all kinds
of practice at variance with those laws must be
bad. Since, from the very nature of things, this
cannot be otherwise, it follows that, by a careful
consideration and due understanding of these laws,
the intelligent cultivator will acquire the most
certain means of improving his practice.

BOOK I.

OF THE PRINCIPAL CIRCUMSTANCES CONNECTED
WITH VEGETABLE LIFE WHICH ILLUSTRATE
THE OPERATIONS OF GARDENING.

7. A PLANT is a living body composed of an
irritable, elastic, hygrometrical matter, called tissue.
It is fixed to the earth by roots, and it elevates
into the air a stem bearing leaves, flowers, and
fruit. It has no power of motion except when it
is acted upon by wind or other external forces; it
is therefore peculiarly susceptible of injury or benefit from the accidental circumstances that may surround it; and, having no free agency, it is above
all other created beings suited to acknowledge the
power of man.

8. In order to turn this power to account, it is
necessary to study the manner of life which is peculiar to the vegetable kingdom, and to ascertain
what the laws are by which the numerous actions
essential to the existence of a plant are regulated.
It is, moreover, requisite that the causes which
modify those actions, either by increasing or diminishing their force, should be understood.

9. The vital actions of plants have so little re

semblance to those of animals, that we are unable to appreciate their nature in even the smallest degree by a reference to our own sensations, or to any knowledge we may possess of animal functions. Nor, when we have thoroughly studied the phenomena of vegetation, are we able to discover any analogies, except of a general and theoretical nature, between the animal and vegetable kingdoms. It is therefore necessary that plants should be studied by themselves, as an abstract branch of investigation, without attempting to reason as to their habits from what we know of other organic beings; and consequently we are not, in this part of Natural History, to acknowledge any theory which is not founded upon direct experiment, and proved by the most satisfactory course of enquiry.

10. In discussing this subject, it will be most convenient for my present purpose, if I divide the matter into the heads of, 1. Germination ; 2. Growth by the Root; 3. Growth by the Stem ; 4. Action of the Leaves ; 5. Action of the Flowers ; and, 6. Maturation of the Fruit. By this means the life of a plant will be traced through all its principal changes, and it will be easy to introduce into one or other of these heads every point of information that can be interesting to the cultivator; who will be most likely to seek it in connexion with those phenomena he is best acquainted with by their effects.

CHAPTER I.

GERMINATION.

The Nature of a Seed. — Its Duration. — Power of Growth. — Causes of Germination. — Temperature. — Light. — Humidity. — Chemical Changes.

11. A SEED is a living body, separating from its parent, and capable of growing into a new individual of the same species. It is a reproductive fragment, or vital point, containing within itself all the elements of life, which, however, can only be called into action by special circumstances.

12. But while it will with certainty become the same species as that in which it originated, it does not possess the power of reproducing any peculiarities which may have existed in its parent. For instance, the seed of a Green Gage plum will grow into a new individual of the plum species, but it will not produce the peculiar variety called the Green Gage. This latter property is confined to leaf-buds, and seems to be owing to the seed not being specially organised after the exact plan of the branch on which it grew, but merely possessing the first elements of such an organisation, together with an invariable tendency towards a particular kind of developement.

B 4

13. Under fitting circumstances a seed grows; that is to say, the embryo which it contains swells, and bursts through its integuments; it then lengthens, first in a direction downwards, next in an upward direction, thus forming a centre or axis round which other parts are ultimately formed. No known power can overcome this tendency, on the part of the embryo, to elevate one portion in the air, and to bury the other in the earth; but it is an inherent property with which nature has endowed seeds, in order to insure the young parts, when first called into life, each finding itself in the situation most suitable to its existence; that is to say, the root in the earth, the stem in the air.

14. The conditions required to produce germination are, exposure to moisture, and a certain quantity of heat; in addition, it is necessary that a communication with the atmosphere should be provided, if germination is to be maintained in a healthy state. A seed, when fully ripe, contains a larger proportion of carbon than any other living part, and so long as it is thus charged with carbon, it is unable to grow. The only means it possesses of ridding itself of this principle, essential to its preservation, but forming an impediment to its developement as a new plant, is by converting the carbon into carbonic acid; for which purpose a supply of oxygen is necessary. It cannot obtain

oxygen in sufficient quantity from the air, for it is cut off from free communication with the air by various means, either natural, as being enclosed in a thick layer of pulp, or in a hard shell or stone; or artificial, as being buried to a considerable depth below the surface of the soil. It is from the water absorbed in germination that the seed procures the requisite supply of oxygen; fixing hydrogen, the other element of water, in its tissue: and thus it is enabled to form carbonic acid, which it parts with by its respiratory organs, until the proportion of fixed carbon is lowered to the amount suited to its growth into a plant.

15. But the formation and respiration of carbonic acid takes place most freely, though not exclusively, in darkness; if exposed to light, the seed again parts with some of its oxygen, and again fixes its carbon by the decomposition of its carbonic acid.

16. In addition to this, the absorption of water causes all the parts to soften and expand; many of the dry, but soluble, parts to become fluid; sap, or vegetable blood, to be formed; and a sort of circulation to be established, by means of which a communication is maintained between the more remote parts of the embryo.

17. Heat seems to set the vital principle in action, to expand the air contained in the numerous microscopic cavities of the seed, and to produce a

distension of all the organic parts, which thus have
their irritability excited, never again to be destroyed except with death. What degree of heat
seeds find most conducive to their germination,
probably varies in different species. Chickweed
(Alsine media) and Groundsel (Senecio vulgaris)
will germinate at a temperature but little above
32° Fahr.

18. Germination being established, by the absorption and decomposition of water, and by the
requisite elevation of temperature, all the parts
enlarge, and new parts are created, at the expense
of a mucilaginous saccharine secretion which the
germinating seed possesses the power of forming.
With the assistance of this substance, the root,
technically called the radicle, at first a mere point,
or rather rounded cone, extends and pierces the
earth in search of food ; the young stem rises and
unfolds its cotyledons, or rudimentary leaves,
which, if they are exposed to light, decompose
carbonic acid, fix the carbon, become green, and,
by processes hereafter to be explained, when
speaking of leaves, form the matter by which all
the pre-existing parts are solidified. And thus a
plant is born into the world ; its first act having
been to deprive itself of a principle (carbon)
which, in superabundance, prevents its growth ;
but, in some other proportion, is essential to its
existence.

CHAP. II.

GROWTH BY THE ROOT.

Roots lengthen at their Points only. — Absorb at that Part chiefly. — Increase in Diameter like Stems. — Their Origin. — Are feeding Organs. — Without much Power of selecting their Food. — Nature of the latter. — May be poisoned. — Are constantly in Action. — Sometimes poison the Soil in which they grow. — Have no Buds. — But may generate them.

19. THE root, being the organ through which food is conveyed from the earth into the plant, is the part which is the soonest developed. Even in the embryo, at the earliest commencement of germination, it is the part immediately connected with the root that first begins to move, by lengthening all its parts, and protruding itself beyond the seed-coats into the earth.

20. But as soon as this primitive lengthening of the root has taken place, and the upper part of the embryo, namely, the young stem, has begun to exist as a separate organ, the root changes its property, ceases to grow by a general distension of its tissue, and simply increases in length by the addition of new matter to its point. A root is therefore extended much in the same way as an icicle,

by the constant superposition of layer over layer to its youngest extremity, with this difference, however, that an icicle is augmented by the addition of matter from without, while the root lengthens by the perpetual creation of new matter from within.

21. For this reason, the extreme points of the roots are exceedingly delicate, and are injured by very trifling causes; they, moreover, as all newly formed vegetable matter is extremely hygrometrical, have the power of absorbing, with rapidity, any fluid or gaseous matter that may be presented to them. On this account they are usually called *spongelets*.

22. In the roots of ordinary exogens, when the tissue is very young, the spongelet (*fig.* 1. *a*) con-

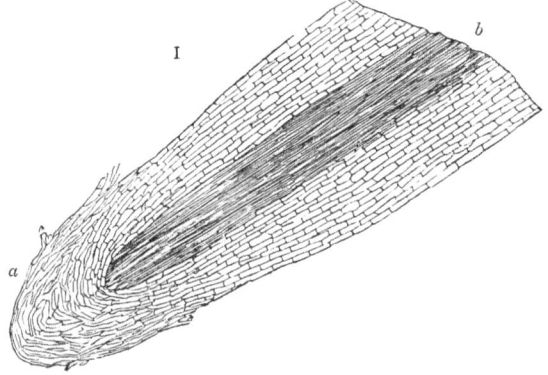

sists of very lax tender cellular tissue, resting upon a blunt cone of woody matter, composed principally of woody tubes, and connected with the alburnum of the stem (*fig.* 1. *b*); it is, therefore, placed in the most favourable position possible for communicating to the general system of circulation the fluids taken up by its highly absorbent tissue.

23. It is the opinion of most vegetable physiologists, that the absorbing or feeding powers of roots are conducted principally at these points; and that the general surface of the root possesses little or no power of the kind. And, indeed, it seems highly probable that this is so, when we consider that the bark of the root, through whose thickness all fluids would have to pass before they reach the alburnum, has at least two offices to perform, either of which might be interfered with by a current of fluid setting through it. One of those offices is to convey in a downward direction, or to store up, the matter which has descended to the roots from the branches and leaves, the other is to give off such superfluous matter as it is necessary for its health that the plant should part with.

24. But although there can be no doubt that the spongelets act as absorbents with more force than any other part of the root, yet it is equally certain that the whole surface of young roots also possesses an absorbing property, only in a more limited degree. It is not until their tissue is solidified that roots become incapable of passing fluid through their sides; and when very young

and soft, there is probably but little difference
between their action and that of the spongelets
themselves; for it is to be remembered that the
latter are not special organs, but are only the very
youngest part of the root.

25. The absorbent power of the spongioles
must be much greater than would have been supposed, if we consider that it is almost entirely
through their action that the enormous waste of
fluid, which takes place in plants by perspiration,
is made good; and hence their importance to
plants, and the danger of destroying them, become
manifest.

26. The spongioles and youngests parts of roots
are found to be rich in nitrogen, a principle once
supposed to be unknown in the vegetable kingdom; and it seems that a supply of this gas is indispensable to their healthy condition.

27. Roots being furnished with the power of
perpetually adding new living matter to their
points, are thus enabled to pierce the solid earth in
which they grow, to insinuate themselves between
the most minute crevices, and to pass on from place
to place as fast as the food in contact with them is
consumed.   So that plants, although not loco-
motive like animals, do perpetually shift their
mouths in search of fresh pasturage, although their
bodies remain stationary.

28. The only known exceptions to the rule that

roots do not lengthen by a general distension of their tissue, occur in parts growing in air or water, which are non-resisting media, or in certain endogenous trees, whose roots lengthen to such a degree as to hoist the trunk up into the air off the ground, with which it at first was level.

29. It is not, however, merely in length that the root increases; if such were the case, all roots would be mere threads. They also augment in diameter, simultaneously with the stem, and under the influence of exactly the same causes. Neither is it by an embryo alone that roots are formed. A plant, once in a state of growth, has the power of producing roots from various parts, especially from its stem, and from older roots.

30. The immediate cause of the formation of roots is involved in obscurity, and is one of the most important parts of vegetable physiology still to be investigated with reference to horticulture. We all know how difficult it is to cause the cuttings of some kinds of plants to produce young roots, and how rapidly they are emitted by others; it is to be supposed, that the difficulty would be diminished in all such cases, if we knew exactly under what circumstances roots are formed. Nothing, however, sufficiently certain and general to merit quotation has yet been ascertained concerning this important subject, except the following facts, viz. that roots are most readily, if not

exclusively, formed in darkness and moderate
moisture ; that they are not, like branches, the
developement of previously formed buds, but appear fortuitously and irregularly from the woody
rather than the cellular part of a plant; and that
their production is in some way connected with
the presence of leaves or leaf-buds, because portions
of a stem having neither leaves nor leaf-buds produce roots unwillingly, if at all; and that such
roots perish if their appearance be not speedily
followed by the formation of leaves. Thus, although the first appearance of the root in the
embryo plant, at the time of germination, precedes
the expansion of the seed-leaves, yet the young
root will not live unless the seed-leaves are enabled
to act.

31. But, although the immediate cause of the
formation of roots is unknown, the remote cause is
apparently the elaboration of organisable matter by
the leaves; for there can be no doubt that the
developement of roots is much assisted by the descending sap. When a ring of bark is removed
from a branch, if the wound is wrapped in damp
moss, roots will invariably push from the upper lip
of the wound, while the lower will produce none ;
a fact so well known, that it has been one of the
causes of an opinion, that roots are bundles of
wood liberated from the central perpendicular

system, and that the wood itself is nothing but a mass of roots formed by the leaves and buds.

32. The principal office of the root is to attract food from the ground. For this purpose it is furnished, as has been seen, with an extremely hygrometrical point or spongelet, which is capable of absorbing incessantly whatever matter of a suitable kind may lie in its neighbourhood. Its force of absorption is always proportioned to the quantity of food that a plant requires : when the sap is consumed rapidly by the leaves, as in the spring, the roots are in rapid action also ; and as the summer advances, and leaves require a smaller quantity of food, the roots become more and more torpid.

33. The proportion borne by the root to the stem is very variable. In such plants as succulent Euphorbias, and probably in all plants whose perspiring powers are feeble, the roots are much smaller than the stem ; but, in others, the surface of these organs must be very much greater than that of the stem and leaves. In young Oaks this is well known to be the case, but the disproportion diminishes as such plants advance in age.

34. There is no period of the year when the roots become altogether inactive, except when they are actually frozen. At all other times, during the winter, they are perpetually attracting food from the earth, and conveying it into the interior of the plant, where it, at that season, is stored up till it is

c

required by the young shoots of the succeeding
year. The whole tissue of a plant will therefore
become distended with fluid food by the return of
spring, and the degree of distension will be in proportion to the mildness and length of the previous
winter. As the new shoots of spring are vigorous
or feeble in proportion to the quantity of food that
may be prepared for them, it follows, that the
longer the period of rest from growth, the more
vigorous the vegetation of a plant will become
when once renewed, if that period is not excessively protracted.

35. Powerful as the absorbing action of roots is
found to be, those organs have little or no power of
selecting their food ; but they appear, in most
cases, to take up whatever is presented to them in
a sufficiently attenuated form. Their feeding property depends upon the mere hygrometrical force
of their tissue, set in action in a peculiar manner
by the vital principle ; this force must be supposed
to depend upon the action of capillary tubes, of
which every part of a vegetable membrane must,
of necessity, consist, although they are, in all cases,
invisible to the eye, even aided by the most powerful microscopes. Whatever matter is presented to
such a set of tubes will, we must suppose, be attracted through them, provided its molecules are
sufficiently minute ; and, as we have no reason to
believe that there is, in general, any difference in
the size of the molecules of either gaseous matter or

fluids consisting principally of water, it will follow that one form of such matters will be absorbed by the roots of plants as readily as another. For this reason, plants are peculiarly liable to injury from the presence of deleterious matter in the earth ; and it is probable that, if in many cases they reject it, it is because it does not acquire a sufficient state of tenuity ; as in the case of certain coloured infusions.

36. But, although this appears to be a general rule, there are some exceptions of importance. If a Pea and a grain of Wheat are placed side by side in earth of the same kind, and made to grow under the same circumstances, the Wheat plant will absorb silex in solution from the earth, and the Pea will absorb none ; whence it would seem that the Pea is unable to receive a solution of flint into its system, and that, consequently, it possesses what amounts, practically, to a power of selection. In like manner, Dr. Daubeny has proved that Pelargoniums, Barley, and the Winged Pea (Tetragonolobus) will not receive strontian ; and it is mentioned by Saussure, that he could not make Polygonum Persicaria absorb, by its roots, a solution of acetate of lime, although it took up muriate of soda (common salt) freely.

37. It is a curious fact that the poisonous substances which are fatal to man are equally so to plants, and in nearly the same way. So that, by

presenting opium or arsenic, or any metallic or alkaline poison, to its roots, a tree may be destroyed as readily as a human being.

38. The natural food of plants consists of carbon in the state of carbonic acid, of nitrogen, certain earths and salts, and water. The latter, if distilled, has little power, by itself, of sustaining vegetable life : but, as in nature it is universally mixed with various other substances, it conveys to the roots the organisable matters that are required ; and it furnishes, by its decomposition, a considerable supply of the oxygen consumed in the formation of carbonic acid, and all the hydrogen that is incorporated in the tissue of plants. It has been proved, experimentally, that plants cannot long exist upon pure water ; but, if they are so circum-stanced as to be able to obtain and decompose carbonic acid, they will grow in the absence of other matters. It is only, however, when the peculiar principles, whether earthy or saline, on which they naturally feed, are presented to them, that they become perfectly healthy ; and especially when they have the means of obtaining nitrogen, which appears, from its great abundance in the youngest parts, to be indispensable to plants upon the first formation of their tissue. *

* Mr. Rigg states that those seeds of the same kind, which contain the largest quantity of nitrogen, germinate the earliest. He found nitrogen in young roots having the proportion of one

39. In addition to their feeding properties *, roots are the organs by which plants rid themselves of the secreted matter which is either superfluous or deleterious to them. If you place a plant of Succory in water, it will be found that the roots will, by degrees, render the water bitter, as if opium had been mixed with it ; a Spurge will render it acrid ; and a leguminous plant mucilaginous. And, if you poison one half of the roots of any plant, the other half will throw the poison off again from the system. Hence it follows, that, if roots are so circumstanced that they cannot constantly advance into fresh soil, they will, by degrees, be surrounded by their own excrementitious secretions.

40. It would also seem to follow that, under the circumstances just named, they would be poisoned, because they have little power of refusing to take up whatever matter is presented to them in a fitting state (35.). But it is by no means certain that the excrementitious matter of all plants is poisonous either to themselves or to others ; and therefore the consequences of roots growing in soil from which

---

to five of carbon. Theodore de Sausure also ascertained that germinating seeds absorb this gas.

* According to Mr. Knight, the roots of trees retain the original vigour of the variety, after the trunks have become debilitated ; or, to use his own words, the powers of life do not become expended so soon in roots as in bearing branches. (See *Hort. Trans.*, vol. ii. p. 252.)

they cannot advance are uncertain, and only to be judged of by actual enquiry into the nature of the secretions.

41. In general, roots have no buds, and are, therefore, incapable of multiplying the plant to which they belong. But it constantly happens, in some species, that they have the power of forming what are called adventitious buds; and, in such cases, they may be employed for purposes of propagation. There is no rule by which the power of a plant to generate such buds by its roots can be judged of; experiment is therefore necessary, in all cases, to determine the point.

CHAP. III.

GROWTH BY THE STEM.

Origin of the Stem. — The growing Point. — Production of Wood, Bark, Pith, Medullary Rays. — Properties of Sapwood, Heart-wood, Liber, Rind, &c. — Nature and Office of Leaf-buds. — Embryo-buds. — Bulbs. — Conveyance of Sap, and its Nature.

42. As soon as the root is fully in action, which is shortly after it has begun to lengthen, the vitality of the living point that exists at the bottom of the seed leaves is excited, and a stem begins to be formed. At first the stem is a mere point of

living matter, often invisible to the eye, but sometimes partially developed ; in which latter case it
is called the plumule. But, as soon as nutritive
matter is conveyed into it by the nascent root, all
its parts receive an impulse, which forces them
into a growth upwards ; what matter already exists
is distended, enlarged, and solidified ; new matter
is rapidly generated in all directions from the vital
centre, and, if it were not for the current setting
upwards from the root, it would possibly grow
into a spherical figure. Pressed upon however
by the surrounding earth, impelled upwards by
the current of sap ascending from the root, and
attracted into the air by the necessity it feels of
respiration, the young stem assumes a cylindrical
form, its sides having a tendency to solidify, and
its point to grow longer. This point, or plumule,
or first leaf-bud, soon attracts to itself the food
which the root procures from the earth, and a
part of the nutritive matter which is stored up in
the seed leaves. It feeds especially upon the
latter until the store is exhausted, and by the
time this happens it is clothed with leaves which
are themselves able to feed it after the seed leaves
have perished. In brief, the stem is a branch produced by the first leaf-bud which the embryo plant
possesses.

43. When the stem is first called into existence,
it is merely a small portion of cellular tissue : an

organic substance, possessing neither strength nor
tenacity, and altogether unsuited to the purposes
for which the stem is destined. If such matter
formed exclusively its solid contents, the stem
would have neither toughness nor strength, but
would be brittle like a mushroom, or like those
parts of plants of which cellular tissue is the exclusive component, such for example as the club-
shaped spadix of an Arum, or the soft prickles of
a young Rose branch. Nature, however, from the
first moment that the rudiment of a leaf appears
upon the growing point of a stem, occupies her-
self with the formation of woody matter, consisting of tough tubes of extreme fineness, which
take their rise in the leaves, and which, thence
passing downwards through the cellular tissue, are
incorporated with the latter, to which they give
the necessary degree of strength and flexibility.
In trees and shrubs, they combine intimately with
each other, and so form what is properly called
the wood and inner bark ; in herbaceous and annual
plants, they constitute a lax fibrous matter. No
woody matter appears till the first leaf, or the seed
leaves, have begun to act; it always arises from
their bases ; it is abundant, or the contrary, in proportion to the strength, number, and developement of the leaves ; and in their absence is absent
also.

44. When woody matter is first plunged into

the cellular tissue of the nascent stem, it forms a
circle a little within the circumference of the stem,
whose interior it thus separates into two parts :
namely, the bark or the superficial, and the pith
or the central, portion ; or, in what are called Endogens, into a superficial coating analogous to
bark, and a central confused mass of wood and
pith intermingled.   The effect of this, in Exogens,
is, to divide the interior of a perennial stem into
three parts, the pith, the wood, and the bark.

45. As the cellular tissue of the stem is not
sensibly lengthened more in one direction than
in another, and as it is the only kind of organic
matter that, in stems, increases laterally, it is sometimes convenient to speak of it under the name of
the *horizontal system* ; and, for a similar reason, to
designate the woody tubes which are plunged
among it, and which only increase by addition of
new tubes having the same direction as themselves,
as the *perpendicular system.*

46. Wood properly so called, and liber or
inner bark, consist, in Exogens, of the perpendicular system, for the most part ; while the pith
and external rind or bark are chiefly formed of
the horizontal system.   The two latter are connected by cellular tissue, which, when it is pressed
into thin plates by the woody tubes that pass
through it, acquires the name of medullary rays.
It is important, for the due explanation of certain

phenomena connected with cultivation, to understand this point correctly ; and to remember that,
while the perpendicular system is distributed
through the wood and bark, the horizontal system
consists of pith, outer bark, and the medullary
processes which connect these two in Exogens,
and of irregular cellular tissue analogous to medullary rays in Endogens.   So that the stem of a
plant is not inaptly compared to a piece of linen,
the horizontal cellular system representing the
woof, and the woody system the warp.

47. Whenever the
stem is wounded, the
injury is repaired by
the cellular or hori-
zontal system, which
forms granulations that
eventually coalesce in-
to masses (*fig.* 2. A),
within which the per-
pendicular system or
woody matter is sub-
sequently developed.
Thus the restoration
of the communication
between the two sides
of an annular excision
is effected by granu-

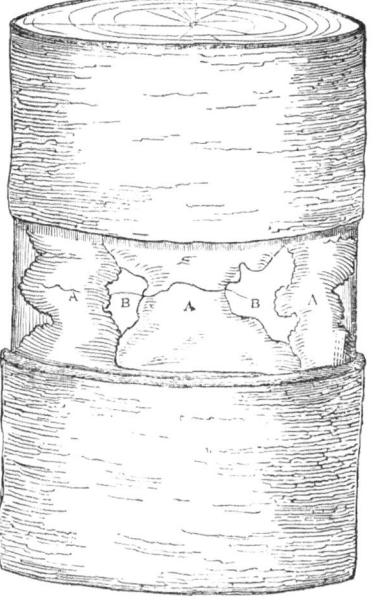

lations of the upper and lower lips, and of the

medullary rays, which finally run together over the wood (*fig.* 2. B), and form a coating below which new liber and alburnum may be generated. In cuttings, the "callus," which forms at the end placed in the ground, is the cellular horizontal system, preparing for the reception of the perpendicular system, which is to pass downwards in the form of roots. Many plants will endure extensive lacerations of their surface, and close up such wounds with great facility. The well known fact of large inscriptions cut in trees below the bark (which inscriptions were effected by removing very broad spaces of the bark and wood) being covered over in time by new bark and wood, so as to be no longer visible from the outside, sufficiently prove this. In such cases, however, the reparation of the injury takes place chiefly, if not exclusively, by the annual addition of new matter to the lips only of the wound, the effect of which is to reduce the circle annually to a less diameter, till at last the centre is closed up.

48. In the bark of trees and shrubs, two distinct parts are found: the one external and cellular; and the other internal, resting upon the wood, and consisting of woody matter mixed with cellular. The external is the rind or cortical integument, the internal is the liber. These two parts grow independently of each other, by their inner faces; the rind belonging exclusively to the

horizontal system, the liber composed of the perpendicular and horizontal systems intermixed.

49. In all Exogenous plants whose stems acquire
an age beyond that of a very few years, the wood
is distinguishable into two parts, heart-wood, and
sap-wood or alburnum. The former is more or less
central, and coloured brown or some dark tint;
the latter is external, pale yellow, and much softer.
Heart-wood was originally alburnum, and altered
its nature with age, in consequence of the solid
matter with which all its tubes and vessels were
choked up; alburnum is the youngest wood, with
all its communications free and open, no solid
matter having had time to accumulate within them.
The reason why solid matter collects in the tubes
of wood, so as gradually to choke them up, is
this: the wood is the channel through which
all the fluid matter of a plant, whether crude or
digested, passes, in its way upwards to the leaves,
or in its horizontal direction from the bark to the
central parts of the stem. When sap leaves the
earth and passes into the stem, it ascends by the
woody matter of the finest fibres of the root;
having left them, it flows into the new wood from
which those fibres emanated, and passes along this
until it reaches the leaves; on its return from
them it descends through the liber, in part passing
off horizontally towards the centre through the
medullary rays. Wherever it passes it deposits a

portion of its solid parts; and, consequently, that
portion of the wood, namely, the oldest or the
heart-wood, through which it has passed the most
frequently, will have the greatest quantity of
matter accumulated within it, independently of
all other reasons for its hardening.

50. The stem of a plant consists, then, of the
following parts, viz. :   1. *Wood*, the oldest of
which is heart-wood, and the newest alburnum ; and
this is the substance through which sap ascends :
2. *Bark*, the external coating, down the liber
or inner face of which sap descends : 3. *Pith*, a
central portion of the horizontal system : and, 4.,
Medullary Rays, serving to connect the rind with
the pith, to hold all the parts together, and to
maintain a communication between the centre and
the circumference of a stem.   The stems of all
plants have these four parts more or less evident.
They are most visible in European trees or shrubs,
in any of which they can be distinctly observed ;
they are least apparent in annual and herbaceous
plants, because their lines of separation are not de-
fined, all the four parts adhering to each other so
firmly as to render it difficult to separate them ;
and in Endogens they are all mixed together, in
consequence of the manner of growth of those
plants not requiring the same kind of arrangement
of parts as is indispensable in Exogens.*   This will

* As this work excludes everything botanical that does

be sufficiently illustrated by the comparison of the stems of an Oak, a Cabbage, and an Asparagus.

51. Tubers, the root-stock of the Iris and Ginger, what are called the roots (cormi) of the Colchicum and Crocus, are all so many different forms of stem.

52. It is the property of a stem, during its growth, to form upon its surface, at irregularly increasing or diminishing distances, minute vital points of the same nature as that in which the stem itself originated. Each of those points becomes, or may become, a leaf-bud, capable of forming other stems or branches like that on which it appeared ; and each is protected and nourished by

---

not directly bear upon horticultural purposes, I have not explained the difference between Exogens and Endogens ; wishing the reader to refer for information upon all such points to works upon pure botany. Nevertheless, as these words are of frequent occurrence, I may as well state that they denominate the two greatest classes in the vegetable kingdom, to one or other of which almost all the flowering plants of common occurrence are referable, and that they derive their names from the peculiarity of their manner of growth. EXOGENS (literally *outside-growers*) are plants whose woody matter is augmented annually by external additions below the liber ; and, consequently, they are continually enclosing within their centre the woody substances formed in previous years ; to such plants, a lateral communication between the centre and the circumference, by means of medullary rays, seems necessary. ENDOGENS (literally *inside-growers*) are plants whose woody matter is augmented annually by internal additions to their centre ; and, consequently, they are continually pushing to their circumference the woody substance formed in previous years.

a leaf which springs from the bark immediately below the bud. Such leaf-buds are the parts that enable a stem, when reduced to the state of a cutting, to produce a new individual like itself; and, without them, no propagation by portions of the stem could take place.

53. Leaf-buds are capable, under fitting circumstances, of growing when separated from their mother branch, whether they are planted in the earth, or inserted below the bark of a kindred species. In the former case, they emit roots into the soil; in the latter they produce wood, which adheres to the wood on which they may be placed. Under ordinary circumstances, leaf-buds will not form anywhere except at the axils* of leaves; but occasionally they appear from other parts, such as the root, the spaces of the stem which lie between the leaves (the internodes), and even from the leaves themselves. In all such cases, they are termed *adventitious*, because of the uncertainty of their appearance. A very remarkable state of them is the *embryo-bud*, a name applied to the *knaurs*, *knurs*, nodules, or hard concretions, found in the bark of various trees, which seem to have, occasionally, the power of propagating the individual, notwithstanding their deformed and indurated state.

* The *axil* is the acute angle formed by a leaf and stem, at the origin of the former; all bodies growing within that angle are said to be *axillary*.

54. Bulbs are buds of a particular kind, larger than common, containing an unusual quantity of secreted matter, and separable, spontaneously, from the part which bears them. They are magazines in which certain plants store up the nutritive matter collected from the leaves. The identity of a bulb and a bud, in all essential circumstances, is obvious, if the bud of any tree (*fig.* 4.) is compared with the bulbs of the Tiger Lily (*fig.* 3.).

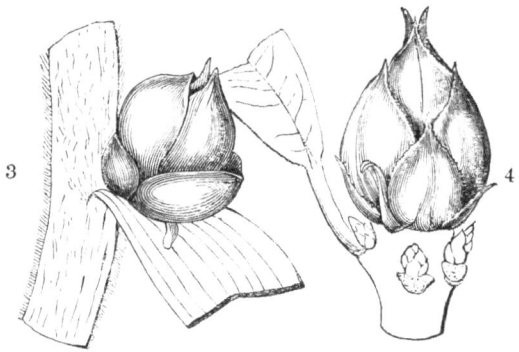

55. As leaf-buds are thus the parents of wood, one of the means of propagating the individual to which they belong, the origin of branches, and consequently the source of the developement of leaves themselves, they may be considered the most important organs of vegetation, so far as any one organ can be called *most* important where all are so mutually dependent the one on the other, and so powerfully concur in maintaining the system of

vegetable life, that it is difficult to abstract one part without impairing the efficiency of the remainder.

56. The office of the stem is, to convey the crude fluid obtained by the roots from the soil, and called *sap*, into the leaves for elaboration, and then to receive it back again. Sap is, originally, water containing various gases, earths, and salts, in solution : but, as soon as it enters the stem, it dissolves the vegetable mucilage it finds there, and becomes denser than it was before ; it is further changed by the decomposition of a part of its water, acquires a saccharine character, and, rising upwards through the alburnum, takes up any soluble matter it passes through. Its specific gravity keeps thus increasing till it reaches the summit of the branches ; and, by degrees, it is all distributed among the leaves. In the leaves it is altered, and then returned into the stem ; not, however, into the alburnum, where it would meet the ascending current, but into the bark, through which it falls, passing off horizontally through the medullary rays into the interior of the stem, and fixing itself in the interior of the bark, especially of the root. It may be said, that, in trees, the alburnum and liber have each two equally important offices to perform: the alburnum giving strength and solidity to the stem, and conveying sap upwards ; the liber not only conveying sap downwards, but covering over the alburnum, protecting it from the air, and enabling it to form

D

without interruption. It is, therefore, indispensable
to the healthy condition of plants, that neither the
alburnum nor the liber should be injured. The
central wood is of little consequence, and may be
destroyed, as it constantly is in hollow trees; and
the rind is of comparatively small importance, for
it is continually perishing under the influence of
the atmosphere : but the liber and alburnum are
naturally in a state of constant renovation, and
cannot be permanently injured without injury to
the plant.

57. But although, under ordinary circumstances,
the sap of exogens rises through the alburnum and
descends through the liber, yet the simplicity of
structure in plants is such, that, together with the
permeability of their tissue, it enables them, in
cases of emergency, to alter their functions, and to
propel their fluids by lateral instead of longitudinal
communications. The trunk of a tree has been
sawed through beyond the pith in four opposite
directions; namely, from north to south, from
west to east, from south to north, and from east to
west, at intervals of a foot, so as completely to cut
off all longitudinal communication between the
upper and lower parts of the stem, as effectually as
if those two parts had been dissevered; and yet
the propulsion of the sap from the roots into the
head of the tree went on as before : which could
only have been effected by a lateral transmission of

this fluid through, or between, the sides of the woody tissue. So when " ringing " is practised, and the alburnum is partially destroyed, the ascending fluid diverges into the stratum of wood beneath the annulation ; and, when it has passed by, it again returns into its accustomed channels ; at the same time, it is probable, although not proved, that some portion of the descending sap forces its way laterally below the wound, out of the bark into the alburnum, using the latter as a means of communicating with the bark below the ring.

Some curious experiments upon this subject were contrived by Mr. N. Niven (*Gardener's Magazine*, vol. xiv.). In one case, he divested the stem of a tree of a deep ring of bark, and of the first twelve lay-ers of wood below it (*fig. 5.*) ; nevertheless the tree continued to live and be healthy From the exposed surface of the wood no sap made its appearance, except from a cut which had been inadvertently made with

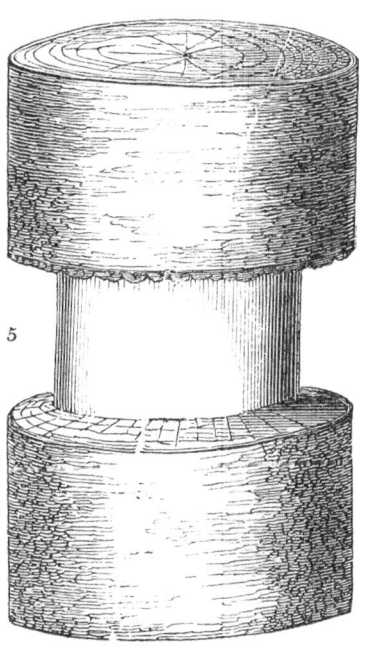

5

the saw on one side, to the depth of, perhaps, five
or six layers of wood beyond the twelve actually
removed.   From that cut a flow of sap took place,
and continued to run during the whole of the season
in which the operation was performed.   In this
case, the sap must have ascended exclusively by
the alburnum.

In another case, by making four deep and wide
incisions into the trunk of a tree (*fig.* 6.), and re-

moving the centre, the upper part of the trunk was
placed upon four separate pillars of bark and albur-

num ; and the tree upon which the operation was performed continued to live for two years, after which it was not observed. In the latter instance, no doubt can be entertained that the whole of the ascending sap was directed into the four pillars of alburnum, which were allowed to remain.

58. The cause of the *flow* of the sap appears to be the attraction of it by the leaves, which continually diminish its quantity ; and the necessity that the sap abstracted should be replaced by a further supply sent upwards from the roots. The consequence of this is, that sap always begins to flow at the ends of branches, a circumstance which has led to the erroneous idea that it proceeds from above downwards through the alburnum. The flow of the sap must not, however, be confounded with the motion of the sap, which takes place in the winter as well as in the summer, and is a mere impletion of the system, caused by the attraction of the roots, unaffected by the exhalation of the leaves.

CHAP. IV.

ACTION OF LEAVES.

Their Nature, Structure, Veins, Epidermis, Stomates. — Effect of Light. — Digestion or Decomposition of Carbonic Acid. — Insensible Perspiration. — Formation of Secretions. — Fall of the Leaf. — Formation of Buds by Leaves.

59. A LEAF is an appendage of the stem of a plant, having one or more leaf-buds in its axil. In those cases where no buds are visible in the axil, they are, nevertheless, present, although latent, and may be brought into developement by favourable circumstances. As this is a universal property of leaves, to which there is no known exception, it follows that all the modifications of leaves, such as scales, hooks, tendrils, &c., and even the floral organs, hereafter to be described, have the same property.

60. Considered with respect to its anatomical structure, a leaf is an expansion of the bark, consisting of cellular substance, among which are distributed veins. The former is an expansion of the rind ; the latter consist of woody matter arising from the neighbourhood of the pith, and from the liber. As the tissue forming veins has a double origin, it is arranged in two layers, united firmly during life, but separable after death, as may be seen in leaves that have been lying for some time in water. Of these layers, one is superior and arises from the neighbourhood of the

pith, the other inferior and arises from the liber; the former maintains a connexion between the wood and leaf; the latter establishes a communication with the bark. As sap, or ascending fluid, rises through the wood, and principally the alburnum, afterwards descending through the liber, it follows from what has been stated, that a leaf is an organ of which the upper system of veins is in communication with the ascending, and the lower system with the descending, current of sap.

61. A leaf has moreover a skin, or epidermis, drawn all over it. This epidermis is often separable, and is composed of an infinite number of minute cavities, originally filled with fluid, but eventually dry and filled with air. In plants growing naturally in damp or shady places it is very thin; in others inhabiting hot, dry, exposed situations, it is very hard and thick; and its texture varies between the two extremes, according to the nature of the species. The epidermis is pierced by numerous invisible pores, called stomates, through which the plant breathes and perspires. Such stomates are generally largest and most abundant in plants which inhabit damp and shady places, and which are able to procure at all times an abundance of liquid food; they are fewest and least active under the opposite conditions. It will be obvious, that, in both these cases, the structure of a leaf is adapted to the peculiar circumstances

under which the plant to which it belongs naturally grows. Now as this structure is capable of being ascertained by actual inspection with a microscope, it follows, as a necessary consequence, that the natural habits of an unknown plant may be judged of with considerable certainty by a microscopical examination of the structure of its epidermis. The rule will evidently be, that plants with a thick epidermis, and only a few small stomates, will be the inhabitants of situations where the air is dry and the supply of liquid food extremely small; while those with a thin epidermis, and a great number of large stomates, will belong to a climate damp and humid; and intermediate degrees of structure will indicate intermediate degrees of atmospherical and terrestrial conditions. It is, however, to be observed, that the relative *size* of stomates is often a more important mark in investigations of this nature than their *number*; those organs being in many plants extremely numerous, but small and apparently capable of action in a very limited degree; while in others, where they are much less numerous, they are large and obviously very active organs. Thus the number of stomates in a square inch of the epidermis of Crinum amabile is estimated at 40,000, and in that of Mesembryanthemum at 70,000, and of an Aloe at 45,000; the first inhabiting the damp ditches of India, the last two natives of the dry rocks of the Cape of Good

Hope : but the stomates of Crinum amabile are among the largest that are known, and those of Mesembryanthemum and Aloe are among the smallest, so that the 70,000 of the former are not equal to 10,000 of the Crinum. Again, the Yucca aloifolia has four times as many stomates as a species of Cotyledon in my collection, but those of the latter are about the $\frac{1}{750}$ of an inch in their longer diameter, large and active, while the stomates of the Yucca are not more than $\frac{1}{2500}$ of an inch long in the aperture, and comparatively inert. The Yucca, therefore, with its numerous stomates, has weaker powers of perspiration and respiration than the Cotyledon.

62. A leaf, then, is an appendage of the stem of a plant, consisting of an expansion of the cellular rind, into which veins are introduced, and enclosed in a skin through which respiration and perspiration take place. It is in reality a natural contrivance for exposing a large surface to the influence of external agents, by whose assistance the crude sap contained in the stem is altered and rendered suitable to the particular wants of the species, and for returning into the general circulation the fluids in their matured condition. In a word, the leaf of a plant is its lungs and stomach, traversed by a system of veins.

63. As the leaf is an extension of the rind of a stem, its epidermis is also an extension of the skin of the same part; and hence it is that in

plants which produce no true leaves, such as the Stapelia, the office of the leaf is performed by the rind and epidermis of the bark.

64. The functions of respiration, perspiration, and digestion, which are the particular offices of leaves, are essential to the health of a plant; its healthiness being in proportion to the degree in which these functions are duly performed. Consequently, whatever tends to impede the free action of leaves, tends also to diminish the healthiness of a plant.

65. These functions are performed by means of the vital forces of vegetation, which we cannot estimate or comprehend, assisted by the influence of an external agent, the nature of whose action may be understood from its effects. That agent is solar light.

66. It is the property of solar light, when striking upon the leaf of a plant, to cause: 1. A decomposition of carbonic acid; 2. An extrication of nitrogen; and, 3., Insensible perspiration. By their vital forces plants appear to decompose water, independently of the action of light.

67. Carbonic acid is originally introduced into the interior of a plant, either dissolved in the water it imbibes by its roots, or by attraction from the atmosphere, or by the combination of the oxygen obtained by a decomposition of water or otherwise, with the carbon in its interior. When a leaf is exposed to the direct influence of the

sun, it gives off oxygen, by decomposing the carbonic acid; whereupon the carbon remains behind
in the interior of the leaf in a solid state. Although the nature of the air thus extricated can
only be determined by a chemist, yet the extrication itself can be easily seen by any one who will
plunge a leaf in water and expose it to the sun;
for bubbles of oxygen will be seen to form themselves upon the surface of the leaf. But, if the
same leaf be observed in the total absence of solar
light, there will be little or no extrication of air,
and what little is given off will be found to be
carbonic acid, which plants exhale at all times in
small quantities; oxygen, however, which was
before expelled, is inhaled. Hence plants decompose carbonic acid during the day, and form it
again during the night, the oxygen they inhale
at that time entering again into combination with
their carbon; and, during the healthy state of a
plant, the decomposition by day, and recomposition
by night, of this gaseous matter, is perpetually
going on. The quantity of carbonic acid decomposed is in proportion to the intensity of the light
which strikes a leaf, the smallest amount being in
shady places; and the healthiness of a plant is,
cæteris paribus, in proportion to the quantity of
carbonic acid decomposed; therefore, the healthiness of a plant should be in proportion to the
quantity of light it receives by day.

68. But, while this is true as a general axiom, it is necessary to observe that some plants are naturally inhabitants of shady situations, and are so organised as to be fit for such places and for no others : plants of this description will not endure full exposure to the sun ; not because an abundant decomposition of carbonic acid is otherwise than favourable to them, but because their epidermis allows the escape of water too freely by insensible perspiration, under the solar stimulus.

69. The mere fact of plants absorbing fluids from the earth, would render it probable that they have some means of parting with a portion of it by their surface ; but that they do perspire is susceptible of direct proof, and is by no means a mere matter of inference.

70. We do not indeed see vapour flying off from the surface of plants ; neither do we from that of animals, except when the air is so cold as to condense the vapour ; yet we know that in both cases perspiration is perpetually going on, and it would appear that in plants it takes place more abundantly than in animals. If a plant covered with leaves is placed under a glass vessel, and exposed to the sun, the sides of the vessel are speedily covered with dew, produced by the condensation of the insensible perspiration of the plant. If the branch of a plant is placed in a bottle of water, and the neck of the bottle is luted to the branch, so that

no evaporation can take place, nevertheless the water will disappear ; and this can only happen from its having been abstracted by the branch which lost it again by insensible perspiration. Hales, an excellent observer, devised many experiments connected with this subject * ; among others the following, which he relates thus : — " August 13. In the very dry year 1723, I dug down $2\frac{1}{2}$ feet deep to the root of a thriving baking pear tree, and laying bare a root half an inch in diameter (*fig. 7.*). I cut off the end of the root at *i,* and put the remaining stump (*i n*) into the glass tube *d r,* which was an inch in diameter, and eight inches long, cementing it fast at *r* ; the lower part of the tube *d z* was eighteen inches long and a quarter of an inch diameter in bore. . . . Then I turned the lower end of the tube (*z*) uppermost, and filled it full of water, and then immediately immersed the small end *z* into the cistern of mercury at the bottom, taking away my finger which stopped up the end of the tube *z.* . . . . The root imbibed the water with so much vigour, that in six minutes' time the mercury was raised up the tube *d z* as high as *z,* namely, eight inches. . . . . The next morning at eight o'clock the mercury was fallen to two inches in height, and two inches of the end of the root *i* were yet immersed in water. As the root imbibed the water, innu-

* See Vegetable Statistics, London, 1727.

merable air bubbles issued out at *i*, which occupied
the upper part of the tube at *r* as the water left

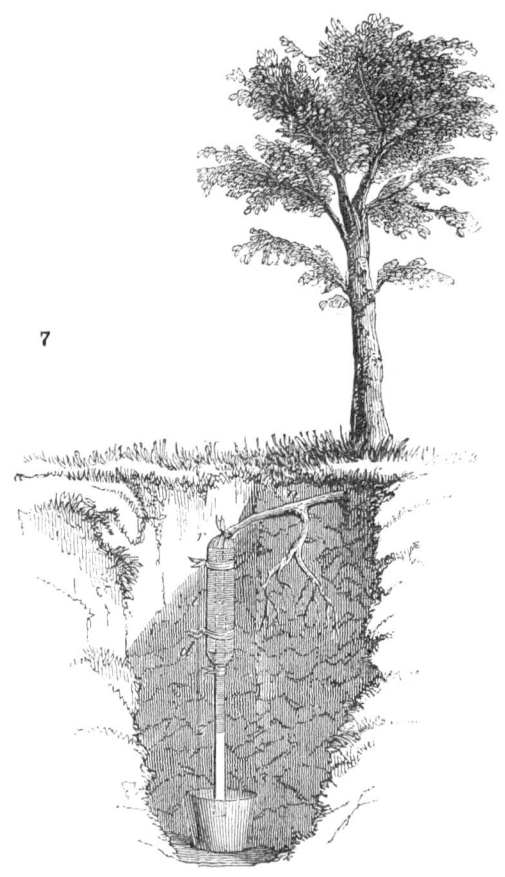

7

it." On another occasion he planted a sunflower
3½ feet high in a garden pot, which he covered
with thin milled lead, cementing all the joints so
that no vapour could escape except through the

sides of the pot and through the plant itself; but providing an aperture, capable of being stopped, through which the earth in the pot could be watered. After fifteen days, viz., from July 3. to August 8., he found, upon making all necessary allowances for waste, that this sunflower plant 3½ feet high, with a surface of 5616 square inches above the ground, had perspired as follows : —

Ounces
Avoirdupois.

In twelve hours of a very dry warm day 30,

On another day -   -   -   - 20,

In a dry warm night without dew - 3,

In a night with some small dew - - 0 ;

and that when the dew was copious, or there was rain during the night, the plant and pot were increased in weight two or three ounces. Other persons have instituted other experiments of a similar nature, the result of all which is, that the insensible perspiration of plants is very considerable.* Hales says his sunflower perspired

* The amount of this force is strikingly illustrated by the following circumstance recorded by the late Mr. Braddick. " One experiment I will mention, as it may serve to show the great power of the rising sap in the vine, while its buds are breaking. On the 20th of March, in the middle of a warm day, I selected a strong seedling vine five years old, which grew in a well prepared soil, against a south-west wall ; I took off its head horizontally with a clean cut, and immediately observed the sap rising rapidly through all the pores of the wood, from the centre to the bark. I wiped away the exuded moisture, and covered

seventeen times more than a man. There is, however, this important peculiarity in vegetable perspiration, that it takes place only or principally in
sunlight. The last experiment shows that, while
the sunflower was losing from twenty to thirty
ounces of water daily during the day, it lost only
three ounces during the night without dew, and
that there was no loss whatever if a slight dew
were present. Here it is probable that the small
amount which was lost at night was parted with
by the sides of the garden pot, and that the plant
itself lost nothing ; for it is in evidence that the
perspiration of plants is in proportion to the
quantity of sunlight that strikes them, and that
in darkness they perspire little or not at all.*   It

---

the wound with a piece of bladder, which I securely fastened
with cement, and a strong binding of waxed twine. The bladder,
although first drawn very close to the top of the shoot, soon
began to stretch, and to rise like a ball over the wound; thus
distended, and filled with the sap of the vine, it felt as hard as
a cricket ball ; and seemed, to all appearance, as if it would
burst.   I caused cold water from a well to be thrown on the
roots of the plant ; but neither this nor any other plan that I
could devise, prevented the sap from flowing, which it continued to do with so much force as to burst the bladder, in
about forty-eight hours after the operation was performed ; the
weather continuing the whole time warm and genial. (*Hort.
Trans.*, v. 202.)

  * M. De Candolle distinguishes between *exhalaison*, or perspiration, which is a vital action, and *deperdition* or evaporation,
which is merely physical.   But the latter is too small in amount
to be worth taking into account for practical purposes.

is no doubt true, that in a dry atmosphere plants will lose their water day and night; but it is equally certain that under such circumstances they will lose very much more by day than by night. They will, however, lose much more by day in a dry atmosphere in a given time, than they will in an atmosphere abounding in moisture.

71. Although perspiration thus appears to be principally excited by the solar rays, and to be in a given plant in proportion to their intensity, yet we are not authorised in concluding that perspiration is not increased or diminished by the medium in which a plant grows. Immersed in water, perspiration is necessarily arrested ; in an ordinary atmosphere, it will be in proportion to the quantity of elastic vapour the atmosphere may contain ; and it is probable, although there are no experiments upon the subject, that it is increased in proportion to the rarefaction of the air.

72. Since a plant does not perspire at night, and since its absorbing points, the roots, remain during that period in contact with the same humid medium as during the day, they will attract fluid into the system of the plant during the night, and, consequently, the weight of the individual will be increased, as Hales found to be the case. In like manner, if plants in the shade are abundantly supplied with moisture at the roots, they also will gain more than they can lose ; and, as this will be a

E

constant action, the result must necessarily be to render all their parts soft and watery.

73. It is evident, from what has been stated, that leaves must derive the food they digest from the earth through the medium of the roots ; and that they, while alive, maintain a kind of perpetual sucking action upon the stem, which is communicated to the spongelets. That this must be of a very powerful nature is apparent from the fact, that the smallest leaf at the extremity of the branch of a lofty tree must assist in setting in action the absorbing power of roots, at a distance equal, perhaps, to three thousand times its own length. If this reciprocal action is not maintained without interruption, and if anything occurs to check it during the period of vegetation, the plant will suffer in proportion to the amount of interruption. For example, if the roots are placed in a warmer medium than the branches, and are thus induced to absorb fluid faster than the slower action of the leaves can consume it, the superfluous sap will burst through the stem and distend its tissue till the excitability is impaired or destroyed. Or if, on the other hand, a branch is caused to grow in a warm medium, while the roots remain in a very cold medium, the former will consume the liquid sap faster than the latter can supply it, and the consequence will be, that the leaves will die, or the fruit will fall off, or the flowers be unable to

set their fruit, from want of a constant and sufficient supply of food. Not that it is necessary for the temperature of the earth and air to be equal, for this does not happen in nature; but it is requisite that they should have some near relation to each other.

74. It is generally, however, believed, that leaves absorb fluid from the air; and their stomates appear well adapted for that purpose, by their position in most abundance on the under side of leaves; and the possibility of recovering drooping or sickly plants, by syringing their epidermis copiously, seems to render this fact almost certain.* It is,

---

* Mr. Knight entertained the opinion, that water is sometimes absorbed by leaves to such an extent as to cause a *descent of the sap through the alburnum;* a derangement of function to which he even ascribed the attacks of mildew fungi upon plants. The secondary and immediate causes, he says, of this disease, and of its congeners, " have long appeared to me to be the want of a sufficient supply of moisture from the soil, with excess of humidity in the air, particularly if the plants be exposed to a temperature below that to which they have been accustomed. If damp and cold weather in July succeed that which has been warm and bright, without the intervention of sufficient rain to moisten the ground to some depth, the wheat crop is generally much injured by mildew. I suspect that in such cases an injurious absorption of moisture, by the leaves and stems of the wheat plants, takes place: and I have proved that under similar circumstances much water will be absorbed by the leaves of trees, and carried downwards through their alburnous substance; though it is certainly through this substance that the sap rises, under other circumstances. If a branch be taken from a tree when its leaves are mature, and one leaf

E 2

however, thought by some, that leaves have no power of absorbing water, even in an elastic state ; and that the renovation of plants by syringing is owing to a diminution of perspiration.

75. It is to the action of leaves, — to the decomposition of their carbonic acid, and of their water ; to the separation of the aqueous particles of the sap from the solid parts that were dissolved in it ; to the deposition thus effected of various earthy and other substances, either introduced into plants, as silex and metallic salts, or formed there, as the vegetable alkaloids ; to the extrication of nitrogen; and, probably, to other causes as yet unknown, — that the formation of the peculiar secretions of plants, of whatever kind, is owing. And this is brought about principally, if not exclusively, by the agency of light. Their green colour becomes intense, in proportion to their exposure to light within certain limits, and feeble, in proportion to their removal from it ; till, in total and continued

---

be kept constantly wet, that leaf will absorb moisture, and supply another leaf below it upon the branch, even though all communication between them through the bark be intersected ; and, if a similar absorption takes place in the straws of wheat, or the stems of other plants, and a retrograde motion of the fluids be produced, I conceive that the ascent of the true sap or organisable matter into the seed-vessels must be retarded, and that it may become the food of the parasitical plants, which then only may grow luxuriant and injurious. " (*Hort. Trans.*, i. 86.)

darkness, they are entirely destitute of green secretion, and become blanched or etiolated. The
same result attends all their other secretions; timber, gum, sugar, acids, starch, oil, resins, odours,
flavours, and all the numberless narcotic, acrid,
aromatic, pungent, astringent, and other principles
derived from the vegetable kingdom, are equally
influenced, as to quantity and quality, by the amount
of light to which the plants producing them have
been exposed.

76. It is, however, to be observed that, as has
already been stated (68.), the capability of plants
to bear the action of direct light varies according
to their specific nature. One species is organised
to suit the atmosphere of a dense wood, into which
diffuse light only will penetrate; another is planted
by nature on the exposed face of a sunburnt rock,
upon which the rays of a shadeless sun are daily
striking: in these cases, the light which is necessary to the one would be destructive of the other.
The organic difference of such species seems to
consist chiefly in the epidermis, which regulates
the amount of perspiration (61.). It is therefore
to be remarked, that it is not the greatest quantity
of light which can be obtained that is most favourable to the healthiness of plants, but the greatest
quantity they will bear without injury. If the
former were true, the concentrated light of a lens
would be better than the strongest ordinary light;

but the effect of the concentrated light of a lens is to burn the surface, and the ordinary solar rays produce the same effect upon many plants, probably by exhausting the tissue of its water faster than it can be supplied from the roots.

77. In the course of time, a leaf becomes incapable of performing its functions; its passages are choked up by the deposit of sedimentary matter; there is no longer a free communication between its parenchyma and that of the rind, or between its veins and the wood and liber. It changes colour, ceases to decompose carbonic acid, absorbs oxygen instead, gets into a morbid condition, and dies : it is then thrown off. This phenomenon, which we call the *fall of the leaf*, is going on the whole year round, except mid-winter, in some plant or other. Those which lose the whole of their leaves at the approach of winter, and are called deciduous, begin, in fact, to cast their leaves within a few weeks after the commencement of their vernal growth ; but the mass of their foliage is not rejected till late in the season. Those, on the other hand, which are named evergreens, part with their leaves much more slowly ; retain them in health at the time when the leaves of other plants are perishing ; and do not cast them till a new spring has commenced, when other trees are leafing, or even later. In the latter class, the functions of the leaves are going on during all

the winter, although languidly; they are constantly attracting sap from the earth through the spongelets, and are, therefore, in a state of slow but continual winter growth. It usually happens that the perspiratory organs of these plants are less active than in deciduous species.

78. In general, a leaf is an organ of digestion and respiration, and nothing more; some leaves have, however, the power of forming leaf-buds, if placed in or upon earth, under suitable circumstances. The Bryophyllum calycinum forms buds at the indentations of its margin; Malaxis paludosa throws off young buds from its margin; Tellima grandiflora occasionally buds at the margins of its leaves: the same thing happens to many Ferns; and several other cases are known.

CHAP. V.

ACTION OF FLOWERS.

Structure of Flowers.—Names of their Parts. — Tendency of the Parts to alter and change into each other, and into Leaves. — Double Flowers. — Analogy of Flowers to Branches.—Cause of the Production of Flowers.—Of Productiveness.—Of Sterility.—Uses of the Parts of a Flower. —Fertilisation. — Hybrids. — Crossbreds.

79. A FLOWER is that part of a plant which is formed for the purpose of reproducing the species

by means of seeds. It consists of floral envelopes and sexes.

80. The floral envelopes are: 1. the calyx, which is usually green, and always the most external; and, 2., the corolla, which is commonly thin, gaily coloured, more fugitive than the calyx, and placed next within it: each of these consists of leaves, called sepals in the calyx, and petals in the corolla. Both calyx and corolla are usually present; but in some cases only one envelope is formed, as in the Marvel of Peru; and in other cases the flower has no envelopes, as in the Willow. Envelopes are, therefore, not a necessary part of a flower.

81. In the middle of the flower stand the sexes, called stamens and pistil, of which the pistil occupies the centre, and the stamens surround it; except in those cases where the sexes are produced in separate flowers, when each sex is central in its own flower. The stamens consist of a filament and an anther, in the inside of the latter of which is secreted a powdery substance called pollen. The pistil consists of ovary, style, and stigma, in the inside of the first of which are ovules or young seeds.

82. Although the floral envelopes may be, and often are, absent, wholly or in part, yet the sexes are always present. Consequently the latter are

all that is essential to a flower, and no part can be a flower from which they are absent.

83. Notwithstanding the difference in form and office of the parts of a flower, they have evidently a strong tendency, in cultivated plants, to change into or assume the appearance of each other. In the Poppy, the Garden Anemone, and many others, the stamens change into petals; in the Anemone, the Ranunculus, &c., the pistil changes into petals; in the Primrose, Cowslip, &c., the calyx changes into petals; in the Houseleek, the stamens become pistils; and so on. Hence the origin of double flowers. In a double Barbadoes Lily, described by me in the *Transactions of the Horticultural Society*, in which the parts were very much confused, the young seeds were borne by the edges of the stamen-like petals. (*fig.* 8.)

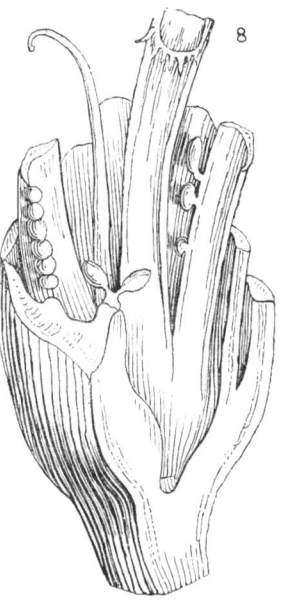

84. In their ordinary state the parts of a flower are extremely unlike leaves, and each has its allotted office, which is not the office of a leaf; they are also incapable of forming leaf-buds in their axils. But, although such is the case, there is

found a strong and general tendency on the parts
of both the floral envelopes and sexes to change

9

to leaves, like the leaves of the stem.   In the
white clover (Trifolium repens, *fig*. 9.), all the
parts often become leaves;
in the Fraxinella (*fig*. 10.)
this has also been remarked * ; so has it in the Nas-
turtium, in Sieversia mon-
tana, and many other in-
stances. A partial alteration
into leaves is of very frequent occurrence in the
parts of a flower. In the

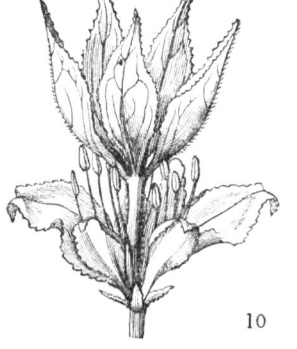

10

Rose, the sepals and pistil are frequently changed
into leaves; in the Double Cherry, the pistil is
almost always to be found in the form of a leaf;

* Proceedings of the Horticultural Society, vol. i. p. 37.

and books on structural botany abound in the records of similar cases. It sometimes happens
that buds are not only formed,
but developed, at the axils of
the parts of a flower, as in
a Celastrus scandens observed
by Kunth. (*fig.* 11.) In the
Pear, it is not uncommon to
find two or three small pears growing out of
an older one (*fig.* 12.), each of which pears may

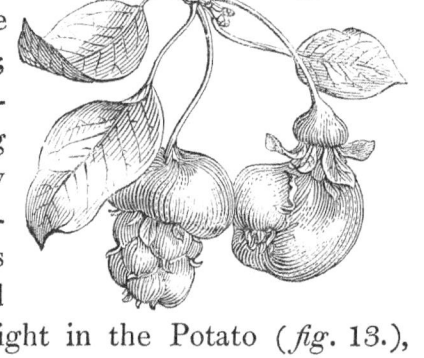

be traced to the axil
of some one of the
parts of the flower;
and rose-buds are fre-
quently seen growing
out of Roses. A very
striking and uncom-
mon case of this
sort was observed
by the late Mr. Knight in the Potato (*fig.* 13.),
whose flowers produced young potatoes in the
axils of the sepals and petals.* Occasionally,
the centre of a flower lengthens and bears its
parts upon its sides, as in the Pear and Apple,
whose fruit is often found in the state of a short
branch. Still more rarely a flower lengthens, and

---

* Proceedings of the Horticultural Society, vol. i. p. 39.
fig. 2.

13

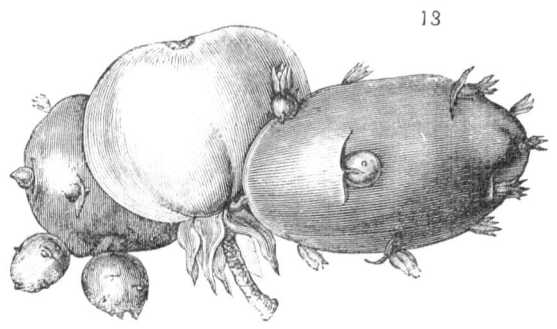

produces from the axils of its parts other flowers
arranged over its sides, as in the Double Pine-apple
of the Indian Archipelago. The following very
striking illustrations of these facts have, among
many others, occurred in the present season
(1839). *Fig.* 14. represents a branch of a Pear
in which one flower (*a*) is in a deformed state,
but still sufficiently recognisable, and another
completely changed into a branch; the calyx
assuming the appearance of leaves or leafy scales
(*s s*), the petals also partially transformed into
leaves (*p p*), while the whole apparatus of stamens
and pistils is converted into an ordinary branch.
Fig. 15. shows the state of plants of Potentilla
nepalensis with their flowers changing to branches:
a is a flower in the ordinary condition; at *b* it is
partly changed in a slight degree; at *c* all the
sepals, petals, and stamens are converted into
leaves, but the pistils are little changed; at *d* the
sepals, petals, and stamens are but little altered,

but the receptacle of the fruit is lengthening into
a branch, and is covered by the carpels partly

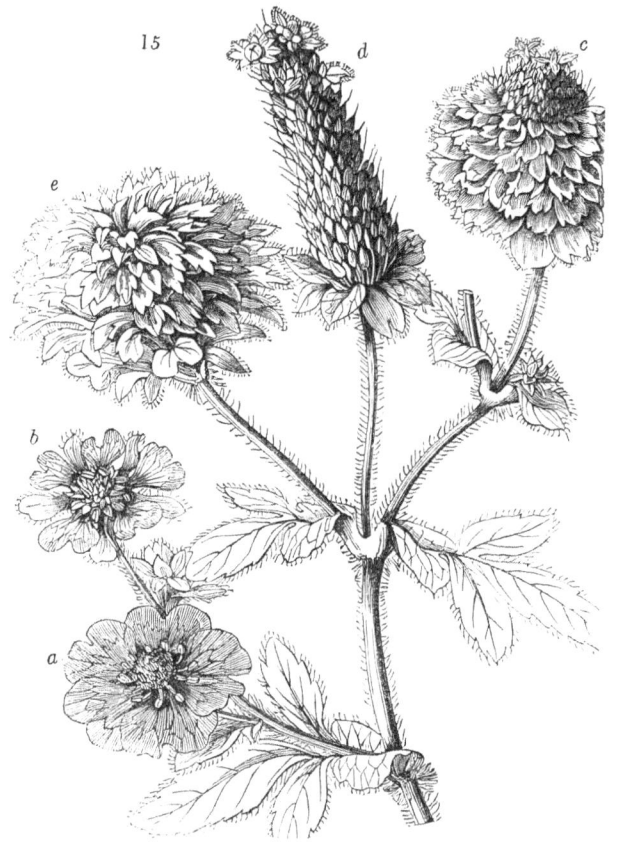

converted into leaves, and some of them near
the apex producing flowers from their axils;
finally, at *e*, the whole of the floral apparatus is
changed into a rosette of leaves. It therefore
appears, that although the parts of a flower are

different both in appearance and office from leaves, yet that they do all assume, under particular circumstances, the same appearance and office. Hence it is inferred that they are really nothing more than leaves in a modified state ; and, consequently, that a flower is a very short branch, and a flower-bud analogous in many respects to a leaf-bud. A leaf-bud is a collection of leaf-scales of the same or similar form, arranged round a central very short branch, having a growing point. A flower-bud is a collection of leaf-scales of different forms, arranged round a central very short branch, not having a growing point under ordinary circumstances. In this latter respect it resembles those buds of the Larch which form leaves in starry clusters, without extending into a branch. Many points in horticulture could not be explained until the existence of this analogy was made out.*

* This doctrine has been taught at different times, by different independent observers. Among other persons, I find that Mr. Knight had come to the same conclusion, at a time when the views of Wolffius and Goethe were quite unknown in England. He says : " The buds of fruit trees which produce blossoms, and those which afford leaves only, in the spring, do not at all differ from each other, in their first stage of organisation, as buds. Each contain the rudiment of leaves only, which are subsequently transformed into the component parts of the blossom, and in some species of the fruit also. I have repeatedly ascertained that a blossom of a Pear or Apple tree contains parts which previously existed as the rudiments of five leaves,

85. What it is that causes a plant to convert some of its buds into flowers, by fashioning the leaves into calyx, corolla, stamens, and pistils, while other buds become branches clothed with ordinary leaves, is beyond the reach of explanation. There are, however, some facts connected with it which require notice. It is clear that plants begin to fructify at some determinate period, varying in different species. In annuals this occurs in a few weeks or months after germination ; in biennials a longer period is required before this condition is arrived at ; and in shrubs and trees a still greater age must be acquired. The American Aloe will not flower before it is thirty years old, under the most favourable conditions ; and, under unfavourable circumstances, the age at which it fructifies is so much increased as to have given rise to the

---

the points of which subsequently form the five segments of the calyx ; and I have often succeeded in obtaining every gradation of monstrosity of form, from five congregated leaves (that is, five leaves united circularly upon an imperfect fruit-stalk) to the perfect blossom of the pear tree. The calyx of the Rose, in some varieties, presents nearly the perfect leaves of the plant, and the large and long leaves of the Medlar appear to account for the length of the segments, in the empalement of its blossom. The calyx of the blossom of the Plum and Peach tree is formed precisely as in the preceding cases, except that the leaves which are transmuted into the calyx separate at the base of the fruit, and become deciduous, instead of passing through and remaining a component part of it." (*Transactions of the Horticultural Society*, vol. ii. p. 364. May 6. 1817.)

vulgar belief that it flowers only after a hundred years. This very curious subject has been little investigated, and we have no comparative statements of the ages at which different species begin to bear; but the fact is certain. It is often, however, in the power of man to advance or retard these periods artificially. Whatever produces excessive vigour in plants is favourable to the formation of leaf-buds, and unfavourable to the production of flower-buds; while, on the other hand, such circumstances as tend to diminish luxuriance, and to check rapid vegetation, without affecting the health of the individual, are more favourable to the production of flower-buds than of leaf-buds. Thus, a plant in a sterile soil and exposed situation flowers sooner and more abundantly than one in a rich and shaded place; young vigorous plants flower later and less abundantly than old ones. In India and China fruit trees are made to bear by cutting their roots, or exposing them periodically to dryness; and in this country the same practice is observed, especially with the fig tree. An apparent exception to this law is found in the fact that a seedling fruit tree may be made, by grafting upon any old stock, to bear flowers at an earlier age than it otherwise would have done; for the effect of grafting it thus is certainly not to render it less vigorous, but the contrary. But it is probable that all these facts arise out of one common

F

law, which is, that the period when a plant begins
to flower depends upon the presence in its system
of a sufficient quantity of secreted matter fit for the
maintenance of the flowers when produced. Under
ordinary circumstances, a considerable part of all
the nutritious secretions elaborated by the leaves
are expended in the production of new leaves;
but, after a time, a greater supply is formed than
the leaves require, and the residue collects in the
system; as soon as this residue has arrived at the
necessary amount, flowers may begin to form. If
the sterile branch of a tree is ringed *, it ceases to
be sterile; and this can only be accounted for
upon the supposition that the secreted matter of
the branch, instead of being conveyed away into
the trunk and roots, is stopped by the annular incision, above which it is compelled to accumulate.
If a tree that is unproductive be transplanted, it
begins to bear; in this case the operation injures

* One of the effects of ringing has been observed to consist
in the formation of numerous barren shoots below the wound,
while fertile shoots appear above it. This is conformable to the
theory of the formation of flowers being determined by a
superabundance of nutritious matter in a given place. The bark
below the annular excision is cut off from a supply of the
sap elaborated by the leaves above it; and, at the same time,
in consequence of the obstruction of the wound to the ascent
of the crude sap, an unusual supply of the latter is forced towards the buds in the bark below the wound, which buds,
being chiefly fed with crude sap, push forth into branches and
leaves, but bear no flowers.

its roots, sap is therefore less abundantly supplied in the succeeding season to the leaves ; the leaves are therefore less able to grow than they previously were, and they consequently do not consume the nutritious matter lying in the branches, and which they would have expended, had they been able to grow with their former vigour ; hence the nutritious matter accumulates, and flower-buds are formed. In this country, if a fruit tree has its crop destroyed one year, it bears the more abundantly the next ; owing, no doubt, to the accumulation in its system of that nutritious matter which would not have been present there, had the crop which was destroyed been allowed to grow : and the reverse of this is well known to be the fact ; an excessive crop one year being followed by a scanty crop the succeeding year. So, when a young seedling fruit tree is made to bear prematurely by grafting it upon an old stock, the effect of which will apparently not be to diminish its vigour, it may be conceived that, in the first place, the seedling will receive a considerable quantity of nutritious matter from the old stock, where it had been already collected, and that thus the supply will be greater than the consumption, however large the latter may be; and, secondly, that, at the time of union of itself with the stock, there will be sufficient interruption of continuity in the bark to oppose some obstacle to the descent from the seedling of whatever matter it may have received or formed. Hence, it is an axiom in vegetable physiology, that the production of flower-buds depends upon the presence of nutritious matter in sufficient abundance for their support.

86. The use of the calyx and corolla is too uncertain and unimportant to demand much notice. The calyx is usually regarded as a protecting organ, and the corolla as a part for the embellishment of the sexes. They neither appear to be of much physiological importance ; more especially not the corolla, or it would not be absent in such large numbers of plants.

87. The use of the stamens is to effect the fertilisation of the young seed contained in the pistil. To this end, the pollen of the anther must be applied to the stigma, the result of which is, that an embryo, the rudiment of a future plant, is generated in the inside of the young seed, and, when mature, is capable of multiplying the species. It is, however, to be observed, that the seed, when ripe, will not renew the species from which it is derived, with all its individual peculiarities ; the seed of a Green Gage Plum, for instance, will not, with any certainty, produce a plant having the sweet green fruit of that variety, but it may produce a plum whose fruit is red and acid. All that the seed will certainly do is to produce a new individual of the plum species ; the peculiarities of individuals are perpe-

tuated by other means, and especially by leaf-buds. (See Book II.)

88. If the pistil of one species be fertilised by the pollen of another species, which may take place in the same genus, or if two distinct varieties of the same species be in like manner intermixed, the seed which results from the operation will be intermediate between its parents, partaking of the qualities of both father and mother. In the first case the progeny is *hybrid*, or mule; in the second it is simply *crossbred*.

89. In general, crossbreds are capable of producing fertile seed, and thus of perpetuating one of the species from which they sprang. Hybrids, on the contrary, are often sterile, and therefore incapable of yielding seed.

90. Reasoning from a few facts, and from the analogy of the higher orders in the animal kingdom, it has been believed that all vegetable hybrids are sterile; and, when sterility is not the consequence of the intermixture of two species, it has been thought that such species are not naturally distinct, however different their appearance. But facts prove that undoubted hybrids *may* be fertile; and when we consider that plants are not analogous to the higher orders of animals, but to the lowest, concerning whose habits we know nothing whatever, it is obvious that no analogical inferences can be safely established.

F 3

CHAP. VI.

OF THE MATURATION OF THE FRUIT.

*Changes it undergoes.— Superior and inferior Fruit.— Is fed
by Branches upon organisable Matter furnished by Leaves.
— Physiological Use of the Fruit. — Nature of Secretions.
—The Changes they undergo. — Effect of Heat —Of Sun-
light — Of Water.—Seeds.—Origin of their Food.—Cause
of their Longevity. — Of their Destruction. — Difference
in their Vigour.*

91. AFTER the fertilisation of the seed has taken
effect, the pistil by itself, or the pistil and surrounding parts, go on growing; alter their
appearance, as well as size; acquire new qualities
of colour, texture, flavour, &c.; and become the
fruit. There are two kinds of fruit essentially very
different : in some instances, the pistil grows separately from the floral envelopes, which drop off, and
the fruit is formed by an enlargement and alteration
of the sides of the pistil only; it is then called
superior : in other instances, the pistil and floral
envelopes all grow together, and the fruit consists
of an enlargement and alteration of the whole
flower; it is then said to be inferior. There is
this essential difference between the two, — that
the superior fruit adheres to the branch by the
base of the pistil alone ; while the attachment of

the inferior fruit is secured by the base, not only of the pistil, but of all the floral envelopes surrounding it. A Peach is a superior fruit; an Apple inferior.

92. A flower being a kind of branch, as has been already shown, and the fruit being an advanced stage of a flower, it follows that a fruit is also a kind of branch. It has originally the same kind of organic connexion with the plant as other branches, and like them requires to be supplied with food, in the absence of which it perishes or languishes. Nevertheless, as its leaves have in but a slight degree the power of forming secretions, and consequently of producing woody tissue in its interior, it will soon drop off its parent, unless the supply of food to it be copious, and its healthy condition permanently secured. Now, as the supply of food to the plant is determined by the attracting force of the leaves of which it consists, and as a superior fruit consists of a smaller number of leaves than an inferior fruit, it follows that the attracting power of an inferior fruit is, *cæteris paribus*, greater than that of a superior, and consequently the former is less likely to drop off; and as the pistil of a superior fruit, being unprotected, is more exposed to external influences, such as that of frost, or a cold dry atmosphere, than an inferior, it also follows that the latter is less liable to suffer from such causes, as

compared with a superior fruit of similar constitutional power. *

93. It is, however, to be remarked that these rules may be interfered with by special causes ; as in the case of the Fig, where the superior fruit is seated on an enlarged receptacle, which acts as if it were a large surface of leaves adhering to the pistil.

94. It may be conceived that, as the fruit is an altered state of a leaf, its physiological action will resemble that of a leaf, in proportion as it retains its organic similitude ; and this is found to happen, a fruit decomposing carbonic acid, &c., under the influence of light, so long as it retains its original green foliaceous character. In the Pea, for example, whose pod is green until it begins to die, the action is always similar to that of a leaf: but in the Peach, whose texture becomes pulpy, and unlike that of a leaf, the physiological action eventually ceases to be exactly that of the latter organ.

* The following table shows which of our commonly cultivated plants have superior or inferior fruits : —

Superior.	Inferior.
Strawberry.	Apple.
Raspberry.	Pear.
Peach.	Quince.
Plum, &c.	Medlar.
Apricot.	Currant.
Cherry.	Gooseberry.
Grape.	Melon.
Fig.	Cucumber.

95. But although a fruit has, like a leaf, the power of forming secretions by elaborating the sap which is attracted into it, yet, because of its small-ness, the amount of this power is inconsiderable: it contributes little to the general secretions of the plant that bears it, but expends its powers in the elaboration of matter for its own use. That it does, however, form wood, like ordinary leaves, is evident, if the flower-stalk of a Cherry is compared with the stalk of the fruit of the same tree ; and this becomes still more apparent when the elaborating forces of many separate fruits are, in consequence of their compact arrangement, brought to contribute to the lignification of a common stalk, as in the Pinaster tree.

96. The great purpose for which the fruit is formed seems to be the protection and nutrition of the seed, the perfect maturation of which is essen-tial to the perpetuation of the races of plants. In most cases the whole of the fluid or nutritious parts is consumed in effecting this end ; but in certain instances there is a surplus, which, if sweet, and unmixed with deleterious secretions, becomes fit for food. In either case, the fruit has, in common with leaves, the power of attracting food from the surrounding parts ; and we see that this property causes the destruction of some fruits by their neighbours which are more advanced in growth, or accidentally more vigorous, and whose attracting

power is so great as to draw to themselves all the food intended for the weaker fruits, which then fall off.   Of the food thus to be consumed in the maturation of the fruit, a portion is derived from the atmosphere, but the principal part has to be prepared by the leaves, which obtain it from the earth through the roots.   It is, therefore, evident, that all causes, of whatever nature, which interfere with the healthy and regular action of leaves and roots, will also interfere with the fruit. Or, if the leaves are placed in such a manner with respect to the fruit, or at so great a distance from it, that the fruit is unable to attract food from them, it must either suffer or perish.   This explains why fruit formed upon naked branches will not continue to grow, and why the presence of a leaf immediately above a fruit, on the same branch, is so beneficial to it.   The size and excellence of fruit will hence be in proportion to the abundance of organisable matter prepared and stored up in its vicinity.*

* The accumulation of sap, and its consequent viscidity, may, however, be attended with disadvantage to a plant, as really happens in the Potato, the most farinaceous varieties of which are liable to a disease called the "curl."   Mr. Knight attributed this to the inspissated state of the sap, which, he conceived, if not sufficiently fluid, might stagnate in and close the fine vessels of the leaf during its growth and extension, and thus occasion the irregular contractions which constitute this disease.   He therefore suffered a quantity of Potatoes, the produce almost

97. Although fruit has, in common with leaves, the property of elaborating the sap, yet there is this difference between them ; that, while leaves return back into the stem what matters they form, fruit retains the principal part of what it forms for the use of itself or of the seeds it contains. This difference is probably to a considerable extent dependent upon the imperfect condition of the bark of the fruit-stalk, which has little power of carrying off from the fruit the matter which is formed within it. In those cases, however, in which the fruit has stomates, the aqueous particles are given off through the surface of the fruit, which then becomes hard or dry when ripe ; but in others, in which there are no stomates, or very few, or very imperfect ones, the aqueous particles cannot be given off to any considerable amount, and the fruit becomes succulent.

98. The maturation of the fruit is dependent, then, upon the action of the leaves and roots, and

wholly of diseased plants, to remain in the heap, where they had been preserved during winter, till each tuber had emitted shoots of three or four inches in length. These were then carefully detached, with their fibrous roots, from the tubers, and were committed to the soil, when, having little to subsist upon except water, not a single curled leaf was produced, though more than nine tenths of the plants which these identical tubers subsequently produced were much diseased. The same effect has been produced by other persons, by taking up the tubers intended for seed before they were full grown, and, consequently, before the excessive inspissation of their secretions had taken place.

the secretions that it forms are principally derived from the former. Consequently, whatever contributes to the healthy condition of the leaves and roots will have a directly beneficial influence upon the fruit, and *vice versâ*. It is, however, certain, that the juices furnished by the leaves undergo a further alteration by the vital forces of the fruit itself, which alteration varies according to species. Thus the fruit of the Peach is sweet, but there is no perceptible sweetness in its leaves; and the fruit of the Fig is sweet and nutritious, while the leaves of that plant are acrid and deleterious.

99. Among the immediate causes of the peculiar changes that occur in the secretions of fruits are heat and light; without which the peculiar qualities of fruits are imperfectly formed, especially in species that are natives of countries enjoying a high summer temperature. It is found that among the effects of a high temperature and an exposure to bright light is the production of sugar and of certain flavours; and that, under opposite circumstances, acidity prevails. As sugar is more rich in carbon than vegetable acids, and has no free oxygen as they have, the sweetness of pulpy fruits ripened under bright sunshine may be understood to arise from the decomposition of carbonic gas, and the expulsion of oxygen, being greater under sunshine than in the shade. Another cause may be, the greater facility with which vegetable acids enter

into combination with gum and starch, and so form sugar, at a high than at a low temperature.*

* Table of the Proportions of Carbon and Water in a few of the commonest Vegetable Secretions.

Substance.	Carbon.	Water, or its Elements.	Oxygen in Excess.
Gum - - (*Berzelius*)	57·318	42·682	
Starch - - (*ditto*)	43·481	56·519	
Tannin - - (*ditto*)	51·160	41·477	3·568
Sugar of Sugar-Cane - (*ditto*)	44·2	55·79	
Grape Sugar (*Th. de Saussure*)	36·71	60·08	3·41
Lignine - - - (*Prout*)	50	50	
Acid, Citric - (*Berzelius*)	41·309	34.234	24·397
—— Malic - - (*Prout*)	40·68	45·76	13·56
—— Oxalic - - (*ditto*)	19·04	42·85	38·11
—— Tartaric - (*ditto*)	32	36	32
—— Ulmic - (*P. Boullay*)	56·7	43·3	
—— Gallic - (*Berzelius*)	56·64	43·36	
—— Acetic - - (*ditto*)	46·23	53·17	
			Hydrogen in Excess.
Oil, Olive - (*Th. de Saussure*)	77·21	10·71	12·08
—— of Almonds (*Henry & Plisson*)	74·40	13·37	4·45
—— of Anise (*Th. de Saussure*)	83·468	14·887	6·465
—— of Lavender (*ditto*)	75·50	14·59	9·55
—— of Roses - (*ditto*)	82·053	4·442	12·631
—— of Turpentine (*GayLussac*)	88·348	- -	11·652
Hydrocyanic Acid (*Gay Lus-sac & Thenard*)	44·39	- -	3·90 (51·71 nitrogen)

The gummy, mucilaginous, and gelatinous parts appear very susceptible of changing into sugar; thus M. Couverchel found that, if Apple jelly is treated with a vegetable acid dissolved in

100. One of the most essential of the alterations which occur in fruits during ripening is, the decomposition or dissipation of the water that they attract from the stem. A diminished supply of water will, under equal circumstances, produce an accelerated maturation, because less time will be required to decompose or dissipate this element; and, on the other hand, an excessive supply of water will retard or prevent ripening, in consequence of the longer time required for the same purpose.

101. Seeds are affected by all circumstances that affect the fruit, which, indeed, as has been already stated, appears to be created for their nutrition and preservation. In general, the fruit attracts organisable matter from the stem through the stalk, and the seed from the fruit through its placenta * ; and this accounts, independently of

---

water, a sugar like that of Grape sugar is the result; that the gum of Peas, placed with oxalic acid in a temperature of 125°, is converted into sugar; that the gum obtained from starch, mixed with the juice of green Grapes, renders it saccharine; and, finally, that tartaric acid, assisted by heat, produces the same effect; which is what causes most fruits to become sweet when cooked. (*De Candolle, Phys. Veg.*, p. 585.)

* The placenta is a soft part of the interior of a fruit, upon which the seed is formed. It is composed of thin-sided parenchyma, the most absorbent of all the forms of tissue, and is in communication, by its whole surface, with the parenchyma of the fruit.

other causes, for the importance of the fruit to
the seed.

102. When the seed is ripe it is dry, all its free
water being parted with ; and its interior is occupied by starch or fixed oil, or some other such
substance, together with earthy matters.   It would
seem that, so long as these secretions remain un-
decomposed, so long does the vitality of the seed
continue unimpaired ; and hence the great age
at which certain kinds of seeds have been found
to grow.*   But, as it is difficult to prevent their
decomposition, so is it difficult to preserve seminal
vitality for any considerable time ; and the differences found in the duration of the growing
powers of seeds probably depend principally upon
chemical differences in their constituent parts.
Oily seeds, which readily decompose, are among
the most perishable ; starchy seeds, which are
least subject to change, are the most tenacious
of life.

103. Warmth, moisture, and an excess of oxygen,
but especially warmth and moisture, while they
are the greatest causes of germination, are pro-

* Not to speak of the doubtful instances of seeds taken from
the Pyramids having germinated, Melons have been known to
grow at the age of 40 years, Kidneybeans at 100, Sensitive-
Plant at 60, Rye at 40 ; and there are now growing, in the garden of the Horticultural Society, Raspberry plants raised from
seeds 1600 or 1700 years old. (See *Introduction to Botany*,
ed. 3. p. 358.)

bably, on that same account, the chief causes of death. Seeds remain dormant so long as the proportion of carbon peculiar to them is undiminished ; water is decomposed by their vital force (14.) ; and its oxygen, combining with the carbon, forms carbonic acid, which is given off. The effect of access of water is, therefore, to rob seeds of their carbon ; and the effect of destroying their carbon is to deprive them of the principal means which they possess of preserving their vitality.

104. Although a seed, if fully formed, is in all cases capable of perpetuating its race, yet there is a difference in the degree to which this capability extends. All seeds will not equally produce vigorous seedlings : but the healthiness of the new plant will correspond with that of the seed from which it sprang. For this reason, it is not sufficient to sow a seed to obtain a given plant : but, in all cases where any importance is attached to the result, the plumpest and heaviest seeds should be selected, if the greatest vigour is required in the seedling ; and feeble or less perfectly formed seeds, when it is desirable to check natural luxuriance. It is apparently for this reason, that old Melon seed is preferred to new ; for the latter would give birth to plants too luxuriant for the small space in which the Melon can be cultivated, under the artificial circumstances required in this country.

105. As both fruit and seeds are maintained at the expense of the leaves, the destruction of the former, when young, will enable the latter to store up against a succeeding season, for the support of future flowers, all that organisable matter which the fruits and seeds destroyed would have otherwise consumed.

CHAP. VII.

OF TEMPERATURE.

Limits of Temperature endurable by Plants. — Effects of a too high Temperature — Of a too low Temperature. — Frost.—Alternations of Temperature.—Day and Night.— Winter and Summer. — Temperature of Earth and Atmosphere.

106. THE extreme limits of temperature which vegetables are capable of bearing, without destruction of their vitality, have not been determined with precision; it is, however, known, that, on the one hand, certain seeds may be boiled without being killed, and that, on the other, they are capable of bearing many degrees of freezing without suffering. In like manner, some plants are found to endure the most intense cold known upon the globe, while others sustain, occasionally, a tem-

perature as high as 140°, as was observed by
Dr. Coulter on the banks of the Rio Colorado.*
The number of plants, however, capable of sustaining such extremes of temperature, is small, and the
greater part of the species known to us are proved
to exist within the limits of 32° and 90°. What
amount of temperature a given species will prefer,
under different circumstances, seems reducible to
no general rule, but has to be determined experimentally in each case, or is judged of by the
known climate of which a plant may be a native.
It is probable that every species has a constitution
better suited to some particular amount of temperature than to any other, although it can bear a
greater or less degree without sustaining injury.

107. Although many plants will live in a temperature much below that of freezing, yet no plant
is able to grow unless the temperature is above 32°,
for physical reasons that require no explanation.
When temperature rises, the air contained in the
minute cells of plants expands, the fluids become
thinner, the excitability of the tissue is aroused,
and, at the same time, insensible perspiration is
commenced, the effect of which is to bring into
play the absorbing powers of the roots, and thus
to set the machinery of vegetation in action. The

* The temperature borne by Oscillatorias in thermal springs
is much higher than this; but no such power is possessed by
cultivable plants.

degree of temperature required to produce this effect is extremely variable in different species of even the same climate, and is, of course, much more variable between plants of different climates. For example, the common weeds called Chickweed, Groundsel, and Poa annua, evidently grow readily at a temperature very near that of 32°; while the nettles, mallows, and other weeds around them, remain torpid. In like manner, while our native trees are suited to bear the low temperature of an English summer, and, in most cases, suffer if they are removed into a country much warmer, such plants as the Mango, the Coffee, &c., inhabitants of tropical countries, soon perish, even in our warmest weather, if exposed to the open air.

108. When, in the case of a given plant, the temperature is permanently maintained at a much higher degree than the species requires, it is over-excited. If the atmosphere is preserved in a proportional state of humidity, the tissue grows faster than the vital forces of the plant are capable of solidifying it, by the decomposition of carbonic acid, and by other means; its excitability is gradually expended, the whole of its organisation becomes enfeebled, the vital functions are deranged, and a state of general debility is brought on.* Such plants are soft and watery, with thin

* According to Mr. Knight, the effect of an excessively high temperature is to cause, in unisexual plants, the production of

leaves, long joints, slender stems, and with no disposition to produce flowers. A slight lowering of
temperature affects them more than a much greater
lowering would have done under other circumstances; and a permanent abstraction of light
readily destroys them. Their inability to decompose carbonic acid, and to assimilate their food in
proportion to their rate of growth, prevents their
becoming so green as is natural to them, and gives
them a pallid hue; and, if it is their property to
secrete other colouring matter, that, like all their
other secretions, is greatly diminished. But, if,
with a preternatural elevation of temperature, there
is a proportionate abstraction of moisture, the loss

male flowers only, while a very low temperature produces the
contrary result.   A Water Melon plant was grown in a house,
the heat of which was sometimes raised to 110° during the
middle of warm and bright days, and which generally varied, in
such days, from 90° to 105°, declining during the evening to
about 80°, and to 70° in the night; the air was kept damp by
copious sprinkling with water, of nearly the temperature of the
external air, and little ventilation was allowed.   The plant,
under these circumstances, grew with great health and luxuriance, and afforded a most abundant blossom; but all its flowers
were male.   " This result," he says, " did not, in any degree,
surprise me; for I had many years previously succeeded, by
long-continued very low temperature, in making Cucumber
plants produce female flowers only; and I entertain but little
doubt that the same fruit-stalks might be made, in this and the
preceding species, to support either male or female flowers in
obedience to external causes." (*Hort. Trans.*, vol. iii. p. 460.)

of fluid, by perspiration and evaporation, goes on faster than the roots can make it good, or the tissue transmit it; the secretions of the species are elaborated faster than the parts to receive them can be formed; the old leaves "burn" and dry up; and young leaves perish, in like manner, as fast as they are formed.

109. Such being the result of preternaturally high temperature in dryness and in moisture, it is easy to conceive that, although such extremes cannot but be prejudicial, yet that they may be approached for particular purposes with advantage. A high temperature and dryness will be favourable to the formation of secretions of whatever kind, while a high temperature, with moisture, will lead to the production of leaves and branches only.*

110. An unnaturally low temperature is productive of evils of another kind. A certain amount of heat is necessary to each particular species, to enable it to grow at all : the immediate effect of heat being to rouse the vital forces, and to bring them into action. If the amount of heat to which a plant is exposed be sufficient to effect this purpose, the functions of the plant are natural and

---

* According to Humboldt, this happens to the Wheat grown about Xalapa in Mexico, which will not mount into ear, but produces an abundance of grass, on which account it is cultivated as a fodder plant.

healthy; the consequences of exceeding it have been explained, those of diminishing it are not less disadvantageous. If the temperature to which a growing plant is exposed is not lowered so much as to destroy it, but just reduced to that point within which it will continue to live, the plant is brought, by the absence of a sufficient exciting cause, into a state not unlike that already described as resulting from over-excitement. It absorbs food from the earth and air, but it cannot assimilate it; its tissue grows, but is not solidified by the incorporation of assimilated matter; aqueous particles accumulate in the interior, a general yellowness ensues, partly from the want of a sufficient power of decomposing carbonic acid, and partly from inability to decompose the water collected in the interior.* The consequence of this is a want of

* The cause of the formation of different colours in different plants is too obscure a subject to suit the purpose of this work. It is, however, as well to observe that the effect of decomposing carbonic acid and exhaling oxygen is the production of a green colour, the intensity of which is, in general, in proportion to the decomposing cause, that is to say, to light: but that, if from any circumstances water is not given off, but is retained in the system and allowed to accumulate, the green colour is altered and changes to yellow; as if the vegetable blue, which must exist in combination with yellow in order to form green, were discharged. Such, indeed, is Macquart's explanation of the phenomenon; and it appears most conformable to theory and fact. For a short explanation of these and other opinions connected with vegetable colouring, see *Introduction to Botany*, ed. 3., book ii. chap. xvi.

the means of forming the usual secretions; flavour, sweetness, nutritive matter, are each diminished; and the power of flowering and fruiting is lost, probably from the absence of a sufficient secretion of organisable matter. (85.) If the unhealthiness of the plant is not so great as to prevent the production of flowers, still they may not expand, as often happens to double roses in cold summers * in England; or, if the flowers do unfold, the fertilising power of the pollen is impaired or destroyed, and no production of seed takes place.

111. Should the temperature be so much lowered as to result in freezing, a destruction of some plants and injury to others take place, owing to physical causes quite different from those whose operation has been explained in the last paragraph. In what degree frost acts upon the vegetable fabric depends upon the specific nature of a plant, the least frost destroying some species, while others, under equal circumstances, endure any known amount of natural cold :· but, as general phenomena, it is in evidence that, when a plant is frozen, the following effects are produced : —

i. The fluids contained within the cells of tissue are congealed, and consequently expanded.

* Want of a sufficiently high temperature, and too much water in the soil, seem to be, either together or separately, the cause of the difficulty experienced by gardeners in making the Double yellow Rose expand its flowers.

ii. Such expansion produces, to some extent, a laceration of the sides of the cells, and impairs excitability by the unnatural extension to which the sides of the cells, if not lacerated, are subjected ;

iii. It expels air from the aeriferous cavities ;

iv. It also introduces air, either expelled from the air passages, or disengaged by the glacial decomposition of water, into parts naturally intended to contain fluid.

v. The green colouring matter and other secretions are decomposed.

vi. The vital fluid, or latex, is destroyed, and the action of its vessels paralysed.

vii. The interior of the tubes, in which fluid is conveyed, is obstructed by a thickening of their sides.

These phenomena may be considered in part mechanical, in part chemical, and in part vital. The two latter are beyond control, and probably depend either upon the quality of fluid and organic matter, which may resist the action of cold in different degrees, according to their various modifications, or else upon specific vitality Salt and water freeze at different temperatures, according to the density of the mixtures, from 4° to 27° ; oil of turpentine at 14° ; oil of bergamot at 23° ; vinegar at 28° ; milk at 30° ; water at 32° ; olive oil at 36° ; oil of anise at 50° ; and it is not to be doubted that, in like manner, the fluid contents of plants, which we know are infinitely modified, will resist the action of cold in very different degrees.* It is recognised, indeed, as a general law, that the difficulty of freezing water is in proportion to its density.

112. The effect of congealing the aqueous particles contained in plants is, in itself, sufficient to cause such a derangement of function as may end in death, and the other supposed causes may be left out of consideration. It will thus follow that, omitting differences arising out of the peculiar nature of different species, plants will suffer from frost in proportion to the abundance and fluidity of their secretions; those whose tissue is driest, and whose secretions are most dense, being the most capable of resisting frost. Hence young shoots are destroyed by a degree of cold which does not affect old shoots of the same species ; and hence, also, the diminished capability of " unripe " shoots, or of plants growing in wet situations, or of trees when they first begin to vegetate, of enduring extreme cold.†

* See a paper on frost in the *Transactions of the Horticultural Society*, new series, vol. ii. p. 308.

† M. DeCandolle gives the following as the laws of temperature with respect to its influence upon vegetation : —

1. All other things being equal, the power of each plant, and of each part of a plant, to resist extremes of temperature, is in the inverse ratio of the quantity of water they contain.

113. The effect of cold is, as has been seen, to
diminish excitability; of heat, to stimulate it :
but, if the latter stimulus were constantly equal,
it may be conceived that the excitability would
soon become impaired or expended. Nature
has however provided against this result, not
only by the fluctuations of temperature that occur
at different periods of the day, but more particularly by the periodical fall of temperature at night
and its rise during the day ; an arrangement intimately connected with all the vital actions of
vegetation. In the day, when light is strongest,
and its evaporating and decomposing powers most
energetic, temperature rises and stimulates the
vitality of plants, so as to meet the demand thus
made upon them ; then, as light diminishes, and
with it the necessity for excessive stimulus, tem-

---

2. The power of plants to resist extremes of temperature is
directly in proportion to the viscidity of their fluids.

3. The power of plants to resist cold is in the inverse ratio of
the rapidity with which their fluids circulate.

4. The liability to freeze, of the fluids contained in plants, is
greater in proportion to the size of the cells.

5. The power of plants to resist extremes of temperature is
in a direct proportion to the quantity of confined air which the
structure of their organs gives them the means of retaining in
the more delicate parts.

6. The power of plants to resist extremes of temperature is in
direct proportion to the capability which the roots possess of
absorbing sap less exposed to the external influence of the
atmosphere and the sun.

perature falls, and reaches its minimum at night, the time when there is the least demand upon the vital forces of vegetation; so that plants, like animals, have their diurnal seasons of action and repose. During the day, the system of a plant is exhausted of fluid by the aqueous exhalations that take place under the influence of sun-light; at night, when little or no perspiration occurs, the waste of the day is made good by the attraction of the roots, and by morning the system is again filled with liquid matter, ready to meet the demand to be made upon it on the ensuing day. No plants will remain in a healthy state unless these conditions be observed.*

114. The alternation of seasons seems to be intended to produce the like effects in a more extended manner ; so that the summer season may be regarded as one long day, and the winter as a night of similar duration. The long days, bright light, and elevated temperature of summer, push the powers of vegetation to their limits ; towards the end of the season excitability becomes impaired, all the vessels and perishable parts are worn out, leaves choke up and can neither breathe nor digest, and the system of a plant, by the incessant exhalation of aqueous matter, becomes

* The incessant vegetation of arctic countries during their summer is an exception to this rule ; but not such as to affect the general truth of the foregoing propositions.

dried up, as it were, and exhausted. At that time, temperature keeps falling, and light diminishing, till at last, upon the arrival of winter, neither the one nor the other is sufficient to excite the vital actions, and a plant sinks into comparative repose. At this time, however, its vital actions are not arrested; if they were, it would be dead or absolutely torpid; they are only diminished in intensity. The roots continue to absorb from the soil food, which is slowly impelled into the system, whence it finds no exit; it therefore gradually accumulates, and in the course of time refills all those parts which the previous summer's expenditure had emptied. In the meanwhile the excitability of the plant is recovered by rest, and may be even conceived to accumulate with the food that the absorbent system of the roots is storing up. At length, when the temperature of the season has reached the requisite amount, excitability is once more aroused, an abundance of liquid food is ready to maintain it, and growth recommences; rapidly or slowly in proportion to the amount of excitement, to the length of previous repose, and to the quantity of food which had been accumulated. In hot climates, where winter is unknown, the requisite periodicity of stimulus and rest is provided for by what are called the dry and the rainy seasons; the former being equivalent to the winter, the latter to the summer, of northern latitudes.

115. As plants have little power of generating heat, like animals, except in particular cases, and very locally *, they are principally dependent upon the media that surround them for the heat which they require. Considering the great importance of heat in their economy, it is, for the purposes of gardening, a most necessary object to ascertain what proportion is usually borne to each other, in different countries, by the temperatures of the earth and atmosphere, the chief media by which plants can be affected. Upon the temperature of the atmosphere there are numerous observations in many countries; upon that of the earth so few as to afford no sufficient data for the solution of this problem. It is usually considered that the temperature of springs affords sufficient evidence of the temperature of the earth ; but, so far as vegetation is concerned, this evidence is unsatisfactory. Springs, deriving their origin from considerable depths, have a nearly uniform temperature all the year round : but the temperature of the earth's surface varies with the seasons ; is extremely different in summer and winter ; and is affected by the quality of the soil, in proportion as that is more or

* Allusion is here, of course, made to the extrication of heat during the periods of flowering and germination, phenomena which have no obvious connexion with cultivation.

less absorbent and retentive of heat.   What we
want to know, as respects vegetation, is, not the
mean temperature of the earth at some distance
from its surface, but the temperature immediately
below the surface ; i. e. of that part of the soil
into which the roots of plants penetrate, and whence
they derive their food.   It is also requisite that
this should be ascertained monthly, so as to furnish
the means of comparing the terrestrial temperature
with the periodical state of vegetation.*   Such
being the case, the temperature indicated by springs

* The following proportions between the mean temperature
of the earth, as indicated by springs, and that of the atmosphere,
have been collected from various sources : —

	Authority.	Temp. of Earth.	Mean Temp. of Atmosphere.
Berlin - -	Wahlenberg	49·28°	46·40°
Carlstrom - -	ditto	47·30	42·03
Upsal - - -	ditto	43·70	42·08
Paris - - - -	(Catacombs)	53·00	51·00
Charlestown - -	Volney	63·00	68·00
Philadelphia - - -	ditto	53·00	53·42
Virginia - -	ditto	57·00	57·00
Massachussets - -	Dewey	47·21	44·73
Vermont -	Volney	44·00	56·00
Raith - - -	Ferguson	47·70	47·00
Gosport - -	Watson	52·46	51·42
Kendal - -	ditto	47·20	47·04
Keswick - - -	ditto	46·60	48·00
Leith - -	ditto	47·30	48·36
South of England - -	Rees' Cyclo.	48·00	50·62
Torrid Zone - - -	Volney	63·00	81·50

will be too high in winter and too low in summer; a most material error.

116. From the observations of Mr. Ferguson, of Raith*, with thermometers buried at different depths in the ground, it appears, that in the years

---

It must be obvious, from these returns, imperfect as they are, that the results are of little value with respect to vegetation, and that indications from springs, from their very nature, can be but little employed in enquiries where it is necessary to determine the fluctuating terrestrial temperature of the surface of the earth in different months. For example, Mr. Ferguson found the temperature of a spring at Raith, 47·7°; but the mean temperature of the earth, one foot below the surface, was 43·58°, and two feet, 44·55°.

* Observations made on the Temperature of the Earth, at One and Two Feet below the Surface, in the Garden of Robert Ferguson, Esq., of Raith.

	1816.		1817.	
	One Foot.	Two Feet.	One Foot.	Two Feet.
January - - -	33·0°	36·3°	35·6°	38·7°
February - -	33·7	36·0	37·0	40·0
March - - -	35·0	36·7	39·4	40·2
April - - -	39·7	38·4	45·0	42·4
May - - -	40·0	43·3	46·8	44·7
June - - -	51·6	50·0	51·1	49·4
July - - - -	54·0	52·5	55·2	55·0
August - - -	50·0	52·5	53·4	53·9
September - -	51·6	51·3	53·0	52·7
October - - -	47·0	49·3	45·7	49·4
November - -	40·8	43·8	41·0	44·7
December -	35·7	40·0	35·9	40·8
Mean of Year -	43·8	44·1	44·9	45·9

1816 and 1817, at that place, in 56° 10′ N. lat.,
and 50 feet above the sea, the mean temperatures,
indicated by geothermometers buried respectively
to the depths of one foot and two feet, varied from
19° to 21° Fahr. between summer and winter, the
earth being colder in winter and hotter in summer
to that amount; and the highest mean observed
was 55·2°, in July 1817, at a foot below the surface.
Other observations, of a similar kind, have been
made in the garden of the Horticultural Society,
from which we learn that, in the valley of the
Thames, the maximum mean of terrestrial temperature, at one foot below the surface, has been
found to be 64·81° in July, which is the hottest
month of the year: but that the greatest difference
between the mean temperature of the earth and
atmosphere is in the month of October, when it
amounted, in the two years during which the observations were made, to between 3 and 4 degrees;
and that, in general, the mean temperature of the
earth, a foot below the surface, is at least one
degree, and more commonly a degree and a half,
above the mean of the atmosphere. In these
cases, if the terrestrial temperatures be compared
with those of the atmosphere, it will be found that
in the spring, when vegetation is first generally
set in motion, the temperature of the earth not
only rises monthly, but retains a mean temperature
higher than that of the atmosphere by from one to

two degrees ; and that, in the autumn, when woody and perennial plants require that their tissue should be solidified, and their secretions condensed, in order to meet the approach of inclement weather, the terrestrial temperature remains higher in proportion than that of the atmosphere, the earth parting with its heat very slowly.*

117. There appears to be no series of direct observations upon the superficial temperature of the earth, at the different periods of vegetation, in other countries ; but some statements are to be found, here and there, concerning the temperature occasionally observed, from which it is to be inferred,

* Quarterly Averages of Temperature obtained from Thermometers buried in the Earth in the Garden of the Horticultural Society ; reduced from the Register kept by Mr Robert Thompson, by Order of the Garden Committee.

	Earth		Mean of Atmosphere.
	One Foot.	Two Feet.	
1837.			
July, August, September - -	62·19°	61·49°	60·44°
October, November, December -	46·13	47·85	43·86
1838.			
January, February, March - -	37·21	38·71	34·57
April, May, June - - -	52·23	50·99	52·01
July, August, September - -	62·15	61·30	60·23
October, November, December -	45·83	47·53	43·28
1839.			
January, February, March -	40·21	41·37	39·51
April, May, June - -	53·05	51·98	52·18
Average monthly mean, from July 1837, to June 1839, inclusive -	49·87	50·15	48·26

H

that the earth is heated, at least for short periods of time, very much above the atmosphere * ; and it is probable that this excessive elevation of temperature is necessary to the healthy condition of many plants. From some interesting observations communicated to me by Sir John Herschel, it appears, that the temperature of the earth, at the Cape of Good Hope, is often excessive. On the 5th of December, 1837, between one and

* Memoranda concerning the Temperature immediately below the Surface of the Earth, occasionally remarked in different Countries.

Egypt	133°—144°	According to Edwards and Colin.
Tropics	Often 126°—134°	Humboldt, *Fragm. As.*
Oronoco	Coarse white sand at 140°, the atmosphere being 84·5°	Humboldt.
France	118°—122°; once 127°, the atmosphere being 91·5°.	Arago, as quoted by Edwards and Colin.
Chile	113°—118° among dry grass.	Boussingault.
New Grenada	85° usual summer temp. 1 foot below surface.	Hay, in *Loudon's Gard. Mag.*, vi. 437.
Cape of Good Hope	159° under the soil of a bulb garden.	Herschel (*MSS.*).
Bermuda	142° thermometer barely covered.	Col. Emmett.
Lantao, China	Water of rice fields 113°; adjacent sandy soil much higher ; for towards midday the black sides of the boat were 142°50.	Meyen.

two o'clock in the day, he observed the heat, under the soil of his bulb garden, to be 159° ; at 3 P.M. it was 150°, and even in shaded places 119° : the temperature of the air in the shade, in the same garden, at the same period, was 98° and 92°. At 5 P.M. the soil of the garden, having been long shaded, was found to have, at 4 inches in depth, a temperature of 102°. "On the 3d of December, a thermometer buried $\frac{1}{4}$ inch deep, in contact with a seedling fir of the year's planting, quite healthy, and having its seed-leaves, marked as follows : — at $11^h\ 25^m$ A. M. 148·2°, at $0^h\ 48^m$ P. M. 149·5°, at $1^h\ 34^m$ P. M. 149·8°, at $1^h\ 54^m$ P. M. 150·8°, and at $2^h\ 46^m$ P. M. 148°." Sir John Herschel observes that such observations " go to show that at the Cape of Good Hope, in the hot months, the roots of bulbous and other plants which do not seek their nourishment very deep, must frequently, and, indeed, habitually, attain temperatures which we can only imitate in our hothouses by actually suspending over the soil plates of red-hot iron. For it must be remarked, that heating the ground *from below* would not distribute the temperature in the same way."

These observations seem to confirm the late Mr. Harvey's suspicions, that the real force of the sun's rays in tropical countries is still far from being ascertained. When, therefore, we are informed by travellers that the temperature in the sun, at

Gondar, has been seen to be 113° (Bruce); at Benares, 110°, 113°, 118° (Harvey); or at Sierra Leone, 138° (Winterbottom); it must be supposed that, in reality, the temperature would have been found much higher in those places, had more efficient means of observation been employed. Mr. Foggo, indeed, succeeded, by means of a large thermometer, having the ball covered with black wool, and fully exposed to the direct rays of the sun, unsheltered from the wind, in obtaining, at Edinburgh, on the 29th of July, at $3^h 10^m$ P.M., an indication of 150°, and at 2^h P.M. of 140°; while another instrument similarly prepared, and resting in contact with herbage, was found to indicate only 119° and 110°; so that, as Mr. Foggo remarks, a difference of 30° was produced in these cases solely from the manner in which the instruments were exposed. (*Edinburgh Philosophical Journal*, No. xxvii.)

118. For horticultural purposes, a very extensive series of observations requires to be made at a very great number of different places, with a view to determine the connexion between the temperature of the soil and the seasons of vegetation; for it does not appear that any such have yet been recorded, except in this country, where, from their fewness, they are by no means so satisfactory as could be desired. In making these, the nature of the soil in which the thermometers are

plunged should, among other circumstances, be very precisely described; for it is obvious that the result will be essentially affected by the peculiar conducting power of the earth.

119. But, although we have no geothermometrical observations which have a direct relation to the connexion between terrestrial temperature and vegetation, yet an approximation to the amount of heat in the earth may perhaps be obtained indirectly. It seems improbable, that the surface of the earth should be colder than the mean temperature of the air that rests upon it; and it seems certain, from the evidence afforded by this country (116.), that, in fact, it is at least a degree or two above it; therefore, in the tropical parts of America, where Humboldt found the mean temperature of the coldest month not to be lower than 79·16° at Cumana, we shall be justified in concluding that the temperature of the earth's surface never falls permanently below that amount; and as the mean summer temperature of the place was found to be 82·04°, so it is probable that the earth will have something above that degree of warmth, on an average, in the summer.*

* For the warmest month, this great observer gives 84·38° as the mean; which corresponds remarkably with the temperature a foot below the surface in New Grenada, where, according to a correspondent of Mr. Hay, it is 85° during summer, " as a gentleman, a planter there, wrote home for his information." (See *Loudon's Gard. Mag.*, vi. 437.)

*** To collect together evidence as to the real amount of temperature at the different seasons of vegetation, in various parts of the globe, would be to render a most important service to horticulture; for it is hopeless to expect that the cultivation of plants can be perfect, in the absence of one of the first data that require to be ascertained. What, for instance, are the terrestrial and atmospheric temperatures of the melon fields of Persia, Bokhara, Spain, or Smyrna, where that delicious fruit acquires its greatest excellence? In the mean while, the few facts recorded in the following table will serve to show the practical importance of such information, it being borne in mind that, as has been already shown (119.), the mean temperature of the soil will probably be, on an average, a degree or two above the recorded means of the warmest and coldest months. Thus, the temperature of the earth at Calcutta, for instance, may be computed to be not more than 88°, nor less than 72°; and if we compare places so similar in climate as Marseilles, Vienna, and London, it will be found that the difference in the terrestrial temperature, as indicated by that of the atmosphere in the warmest month of summer, is quite sufficient to explain why we have so little success in the cultivation of the vine in the open air in England.

120. A Table of Mean Temperatures of the hottest and coldest Months.

	Latitude.	Longitude.	Mean. Temp. of Warmest Month.	Coldest Month.	Authorities.
St. Petersburgh -	59° 56′ N.	30° 19′ E.	65·66°	8·60°	Humboldt.
Moscow - -	55 45 N.	37 32 E.	70·52	6·08	ditto.
Melville Island {	74 47 N.	110 48 W.	39·08	−35·52	Hugh Murray.
	-	-	42·41	−32·19	Ed. Phil. Journal.
Copenhagen -	55 41 N.	12 35 E.	65·66	27·14	Humboldt.
Edinburgh -	55 57 N.	3 10 W.	59·36	38·30	ditto.
Geneva -	46 12 N.	6 8 E.	66·56	34·16	ditto.
Vienna - -	48 12 N.	16 22 E.	70·52	26·60	ditto.
Paris - -	48 50 N.	2 20 E.	65·30	36·14	ditto.
London - -	51 30 N.	0 5 W.	64·40	37·76	ditto.
Philadelphia -	39 56 N.	75 16 W.	77·00	32.72	ditto.
New York -	40 40 N.	73 58 W.	80·70	25·34	ditto.
Pekin - -	39 54 N.	116 27 E.	84·38	24·62	ditto.
Milan - -	45 28 N.	9 11 E.	74·66	36·14	ditto.
Bordeaux -	44 50 N.	0 34 W.	73·04	41·00	ditto.
Marseilles -	43 17 N.	5 22 E.	74·66	44·42	ditto.
Rome -	41 53 N.	12 27 E.	77·00	42·26	ditto.
Funchal -	32 37 N.	16 56 W.	75·56	64·04	ditto.
Algiers -	36 48 N.	3 1 E.	82·76	60·08	ditto.
Cairo -	30 2 N.	30 18 E.	85·82	56·12	ditto.
Vera Cruz -	19 11 N.	96 1 W.	81·86	71·06	ditto.
Havannah -	23 10 N.	82 13 W.	83·84	69·98	ditto.
Cumana - -	10 27 N.	65 15 W.	84·38	79·16	ditto.
Canton - -	23 10 N.	113 13 E.	84·50	57·00	Anglo-Chinese Calendar.
Macao - -	22 10 N.	113 32 E.	86·00	63·50?	ditto.
Canaries -	28 30 N.	16 00 W.	78·90	63·70	Brande's Journal.
Lohooghat(5800 feet above the sea) - -	29 23 N.	79 56 E.	69·34	43·57	{ Trans. Med. Phys. Soc. Calc.
Fattehpúr -	25 56 N.	80 45 E.	74·94	58·74	Gleanings in Science.
Gurrah Warrah -	23 10 N.	79 54 E.	87·45	60·23	ditto.
Calcutta - -{	22 40 N.	88 25 E.	85·70	66·20	ditto.
			86·86	70·10	Journ. As. Soc.
Ava -	21 51 N.	95 98 E.	88·15	64·12	Gleanings in Science.
Bareilly -	28 23 N.	79 23 E.	91·91	56·50	ditto.
Chunar -	25 9 N.	82 54 E.	90·00	58·00	Ed. Ph. Journ.
Cape of Good Hope (Feldhausen) - -	34 23 S.	18 25 E.	74·27	57 48	Herschel (MSS.)
Bahamas - - -	26 30 N.	78 30 W.	83·52	69·07	Hon. J. C. Lees (MSS.)
Swan River -	32 00 S.	115 50 E.	78·00	54·84	Milligan.
Bermuda - - -	32 15 N.	64 30 W.	76·75	57·90	Col. Emmett.

BOOK II.

OF THE PHYSIOLOGICAL PRINCIPLES UPON WHICH THE OPERATIONS OF HORTICULTURE ESSENTIALLY DEPEND.

EVERY operation in horticulture depends for success upon a correct appreciation of the nature of the vital actions described in the last Book; for although there have been many good gardeners entirely unacquainted with the science of vegetable physiology, and although many points of practice have been arrived at altogether accidentally, yet it must be obvious that the power of regulating and modifying knowledge so obtained cannot possibly be possessed, unless the external influences by which plants are affected are clearly understood. Indeed, the enormous difference that exists between the skill of the present race of gardeners and their predecessors can only be ascribed to the general diffusion, that has taken place, of an acquaintance with some of the simpler facts in vegetable physiology.

In attempting to apply the explanations of science to the routine of horticultural practice, it appears desirable, in order to avoid frequent repetition, that all the subordinate details of the art should be omitted, and that those general operations should alone be adverted to which, under many different modifications, and in various forms, constitute the foundation of every gardener's education.

CHAP. I.

OF BOTTOM HEAT.

THIS term is, in common practice, made use of only in those cases where the temperature of the soil in which plants grow is artificially raised considerably above that which we are acquainted with in England; and there seems to be a general idea that such an artificial elevation of temperature is only necessary in a few special instances. It has, however, been shown (116.) that the mean temperature of that part of the soil in which plants grow is universally something higher than that of the air by which they are surrounded, and consequently it appears that nature, in all cases, employs some degree of bottom heat as a stimulus and protection* to vegetation. At the same time, it

---

* That the warmth of the soil acts as a protection to plants may be easily understood. A plant is penetrated in all directions by innumerable microscopic air passages and chambers, so that there is a free communication between its extremities. It may therefore be conceived that if, as necessarily happens,

must be admitted that, in some cases, the amount is extremely small; for Von Baer found Ranunculus nivalis and Oxyria reniformis flowering in Nova Zembla, where the soil was not warmed above $34\frac{1}{2}°$; and, in Jakutzsk, Erdmann states that Summer Wheat, Rye, Cabbages, Turnips, Radishes, and Potatoes are cultivated, although the ground is not thawed above three feet in depth.

That elevating the temperature of moist soil produces an unusual degree of vigour in plants unaccustomed in nature to such an elevation is a fact which requires no proof: it is attested by the condition of vegetation round hot springs, and in places artificially heated by subterraneous fires; and this has probably been the cause of the employment of tan and hotbeds, by which means bottom heat has been generally obtained for rearing delicate species,

---

the air inside the plant is in motion, the effect of warming the air in the roots will be to raise the internal temperature of the whole individual; and the same is true of its fluids. Now, when the temperature of the soil is raised to 150° at noonday by the force of the solar rays, it will retain a considerable part of that warmth during the night : but the temperature of the air may fall to such a degree that the excitability of a plant would be too much and suddenly impaired, if it acquired the coldness of the medium surrounding it; this is prevented, we may suppose, by the warmth communicated to the general system, from the soil, through the roots; so that the lowering of the temperature of the air, by radiation during the night, is unable to affect plants injuriously, in consequence of the antagonist force exercised by the heated soil.

and especially seeds. But if this stimulus acts in the first instance beneficially in all cases alike, it soon becomes a source of mischief in those species which are natives of climates where such terrestrial heat is unknown, the latter "drawing up," as the saying is, becoming weak and sickly, and speedily presenting a diseased appearance (108.).

On the other hand, it is equally well known that, unless the temperature of the soil be raised permanently to at least 75°, the seeds of tropical trees will not germinate; or, if they do, they push forth feebly, and from the first present the sickly appearance of plants suffering from cold (110.). Hence arises the impossibility of making the seeds of tropical plants germinate when sown in the open air in this country, where the mean temperature of the earth seldom rises to 65°, and that for only short periods of time. It is, therefore, obvious that all plants require some bottom heat; but the amount varies with their species, and the only means or power of determining what the amount should be is afforded by the known degree of warmth of the climate of which a plant may be a native.

When plants are cultivated in glass houses, there is little difficulty in supplying them with the amount of bottom heat which they may require; but this can either not be effected at all, or only to a limited degree by a selection of soils and situations, when plants are cultivated in the open air; and hence one of the many difficulties of acclimatising in a cold country the species of a warmer climate.   It is true that plants will exist within wide limits of temperature, and, consequently, a few degrees of difference in the natural bottom heat to which they are exposed may not affect them so far as to destroy them; but it cannot be doubted that the conditions most favourable to their growth are those which embrace a temperature rather above than below that to which they are accustomed in their native haunts.

The Orange tree is found in perfection where the temperature of the soil may be computed to rise to 80° or 85°, and never to fall below 58°, as in the Bermudas, Malta, and Canton.   How injudicious, then, is our practice of exposing it during summer to the open air, in tubs, where the soil scarcely rises in temperature above 66°, and preserving it during winter in cold conservatories, the soil of which often sinks to 36° ; under such circumstances the Orange exists indeed, but where are the perfume and juiciness of its fruit, and where the healthy vigour of its noble foliage?   The Vine cannot be grown in the open air of this country to any useful purpose, except when trained to walls, in soils and situations unusually exposed to the beams of the sun ; it is only then that it can obtain for its roots such a permanent warmth as

75°, which it will have at Bordeaux, or 80° in Madeira.

It may hence be considered an axiom in horti-culture, that *all plants* require the soil, as well as the atmosphere, in which they grow, to correspond in temperature with that of the countries of which they are natives. It has also been already shown, that the mean temperature of the soil should be a degree or two above that of the atmosphere (119.).

This explains why it is that hardy trees, over whose roots earth has been heaped or paving laid, are found to suffer so much, or even to die ; in such cases, the earth in which the roots are grow-ing is constantly much colder than the atmosphere, instead of warmer. We have here*, also, the cause

* Mr. Knight long since mentioned an important fact con-nected with this subject : — " It is well known," he said, " that the bark of Oak trees is usually stripped off in the spring, and that in the same season the bark of other trees may be easily detached from their alburnum, or sap-wood, from which it is, at that season, separated, by the intervention of a mixed cellular and mucilaginous substance ; this is apparently employed in the organisation of a new layer of fibre, or inner bark, the annual formation of which is essential to the growth of the tree. If, at this period, a severe frosty night, or very cold winds, occur, the bark of the trunk, or main stem, of the Oak tree becomes again firmly attached to its alburnum, from which it cannot be sepa-rated till the return of milder weather. Neither the health of the tree, nor its foliage, nor its blossoms, appear to sustain any material injury by this sudden suspension of its functions ; but the crop of acorns invariably fails. The Apple and Pear trees appear to be affected to the same extent by similar degrees of

of the common circumstance of Vines that are forced early not setting their fruit well, when their roots are in the external border and unprotected by artificial means ; and to the same cause is often to be ascribed the shriveling of grapes, which, as we all know, most commonly happens to Vines whose roots are in a cold and unsunned border.

Mr. Reid of Balcarras has, indeed, shown that one of the causes of canker and immature fruit even in orchards is the coldness of the soil. He found that, in a cankered orchard, the roots of the trees had entered the earth to the depth of 3 feet ; and he also ascertained that, during the summer months, the average heat of the soil, at 6 inches below the surface, was 61° ; at 9 inches, 57° ; at 18 inches, 50° ; and at 3 feet, 44°. He took measures to confine the roots to the soil near the surface, and the consequence was, the disappearance of canker, and ripening of the fruit. (*Memoirs of Caledonian Hort. Soc.* vi. part. 2. ; and *Gardener's Magazine*, vii. 55.)

---

cold. Their blossoms, like those of the Oak, unfold perfectly well, and present the most healthy and vigorous character; and their pollen sheds freely. Their fruit, also, appears to set well ; but the whole, or nearly the whole, falls off just at the period when its growth ought to commence. Some varieties of the Apple and Pear are much more capable of bearing unfavourable weather than others, and even the Oak trees present, in this respect, some dissimilarity of constitution." (*Hort. Trans.*, vi. 229.)

If, on the other hand, we take cases of growth in the artificial climate of hot-houses, we find that Bignonia venusta, and many other tropical plants, will not flower unless in a high bottom heat; and that Palm trees, planted in the soil of conservatories which it is impracticable to heat sufficiently, soon become unhealthy.

The reason why it is necessary to plants in a growing state, that the mean temperature of the earth should be higher than that of the air, is sufficiently obvious. Warmth acts as a stimulus to the vital forces (17.), and its operation is in proportion to its amount, within certain limits. If, then, the branches and leaves of a plant are stimulated by warmth to a greater degree than the roots, they will consume the sap of the stem faster than the roots can renew it; and, therefore, nature takes care to provide against this by giving to the roots a medium permanently more stimulating, that is, warmer, than to the branches and leaves.

Such being the fact, it is obvious that one of the first of a gardener's cares should be, to secure the means of insuring a proper temperature to the soil in which he grows his plants, and that this is requisite for hardy as well as tender species; and I entertain little doubt that the time is at hand when it will be considered quite as necessary to furnish heat for the soil as for the air; not, however, heat without moisture, for that would evidently produce much greater evils than it was intended to cure, as has indeed been found by inconsiderate experimenters. I quite agree with Mr. Writgen in believing that it is the temperature and moisture of a soil, much more than its mineralogical quality, that determine its influence upon vegetation. (See *Erster Jahresbericht, &c., am Mittel und Nieder-Rhein*, p. 64.)

Mr. Fintelmann, the king of Prussia's gardener at Potsdam, is celebrated for his success in the difficult art of forcing Cherries, and he has given an account of his practice (*Gard. Mag.*, vol iii., p. 64.), in which it appears that the most peculiar feature is the strict attention he pays to the temperature of the roots. He first soaks the roots in water heated by the mixture of equal parts of boiling and cold water; he afterwards sprinkles the trees with lukewarm water, and he continues to employ it of the same temperature as long as watering is required.

It seems, indeed, clear, that the success of the Dutch in obtaining an abundance of fresh vegetables, such as Lettuces, during the whole winter, is in part owing to their being able to maintain a gentle bottom heat. No doubt this is connected with the abundant light which their forcing structures admit, and with other causes of considerable importance; but none of those causes can be supposed likely, in the absence of the bottom heat, to produce such a result as the Dutch gardeners obtain.

If it is necessary that the temperature of the *soil* in which plants grow should be carefully regulated, and adjusted to their natural habits, it is no less requisite that the *water* in which aquatics are cultivated should be also brought to a fitting heat. Mr. William Kent succeeded well in making many tropical species flower, by growing them in lead cisterns plunged in a tan-bed (*Hort. Trans.*, iii. 34.) in a close heat. In like manner, Mr. Christie Duff procured flowers in abundance from Nymphæa rubra, cærulea, and odorata, by placing them in a cistern in a pine stove upon the end flues, where the fire enters and escapes; or by plunging them into tan-beds in pine houses, varying in temperature from 80° to 100°. (*Hort. Trans.*, vii. 286.) Very lately, Mr. Sylvester, of Chorley, in Lancashire, obtained fine flowers from Nelumbium luteum, by paying attention to the temperature of the water. When he kept the latter at 85°, the plants grew vigorously, and were in perfect health, but flowerless; but by lowering it to 70°—75°, which more nearly approaches the heat to which the plant is naturally accustomed, the magnificent blossoms were produced and succeeded by seeds; the red Nelumbium, however, which inhabits countries with a greater summer heat than the yellow, at the same time suffered by this lowering of temperature, none of its blossom buds having been able to unfold. (*Bot. Mag.*, xiii. n. s. t. 3753.) The water of rice fields, in which the red Nelumbium flourishes, was seen by Meyen at 113° at Lantao, in China (117.).

An opinion has, nevertheless, been entertained, that bottom heat is useless ; there is in the *Horticultural Transactions* (vol. iii. 288.) a paper to show that it is injurious ; and the authority of Mr. Knight has been referred to in support of the opinion, in consequence of that great horticulturist having expressed a belief that the " bark-bed is worse than useless." (*Hort. Trans.*, iv. 73.) But Mr. Knight repeatedly disavowed entertaining any such sentiments. In one place, he stated that the temperature of the air of the stoves in which his Pine-apple and other stove plants grew, *without bark or other hot-bed,* usually varied from 70° to 85° ; and that the mould in his pots, being surrounded by such air, acquired and retained, as it necessarily must, very near the same aggregate temperature, but subject to less extensive variation (*Gard. Mag.*, v. 365.) : in another, he says the temperature of the air was varied in his stove generally from about 70° to 85° of Fahrenheit ; and he ascertained, by keeping a thermometer immersed in the mould of the pots, that the temperature of the soil varied very considerably less than that of the air of the stove ; the mould being in the morning generally some degrees warmer than the air of the house, and in the middle of the day, and early part of the evening, some degrees cooler. (*Hort. Trans.*, vii. 255.)

It is, therefore, clear that he considered a high temperature necessary for the roots of his Pine-apple plants ; and we find, from one of his papers (*Hort. Trans.*, iv. 544.), that he considered it better to obtain the requisite temperature from the atmosphere than from a bark-bed, the usual source of bottom heat, " because its temperature is constantly subject to excess and defect;" and he even admitted that if the bark-bed could be made to give a steady temperature of about 10° below that of the day temperature of the air in the stove, Pine plants would thrive better in a compost of that temperature than in a colder.

It is, therefore, plain that the dispute about bottom heat was not as to the necessity of it, but as to the manner of obtaining it, which, as it concerns the *art* of gardening, I need not further notice.

We have, doubtless, much to learn as to the proper manner of applying bottom heat to plants, and as to the amount they will bear under particular circumstances. It is, in particular, probable that in hot-houses plants will not bear the same quantity of bottom heat as they receive in nature, because we cannot give them the same amount of light and atmospheric warmth ; and it is necessary that we should ascertain experimentally whether it is not a certain proportion between the heat of the air and earth that we must secure, rather than any absolute amount of bottom heat.

It may also be, indeed it no doubt is, requisite to apply a very high degree of heat to some kinds of plants at particular seasons, although a very much lower amount is suitable afterwards; a re-mark that is chiefly applicable to the natives of what are called extreme climates, that is to say, where a very high summer temperature is followed by a very low winter temperature. Such countries are Persia, and many parts of the United States, where the summers are excessively hot, and the winter's cold intense. The seeming impossibility of imitating such conditions artificially will proba-bly account for many of the difficulties we expe-rience in bringing certain fruits, the Newtown pippin, the cherry, the grape, the peach, and the almond, to the perfection they acquire in other countries.

This subject will be frequently recurred to here-after.

CHAP. II.

OF THE MOISTURE OF THE SOIL. — WATERING.

IT has already (38.) been shown that water is one of the most important elements in the food of plants, partly from their having the power of decomposing it, and partly because it is the vehicle through which the soluble matters found in the earth are conveyed into the general system of vegetation. Its importance depends, however, essentially upon its quantity.

We know, on the one hand, that plants will not live in soil which, without being chemically dry, contains so little moisture as to appear dry; and, on the other hand, an excessive quantity of moisture is, in many cases, equally prejudicial. The great points to determine are, the amount which is most congenial to a given species under given circumstances, and the periods of growth when water should be applied or withheld.

When a plant is at rest, that is to say, in the winter of northern countries and the dry season of the tropics, but a small supply of water is required by the soil, because at that time the stems lose but little by perspiration, and consequently the roots demand but little food; nevertheless, some terrestrial moisture is required by plants with

perennial stems, even in their season of rest, because (34.) it is necessary that their system should,
at that time, be replenished with food against the
renewal of active vegetation : hence, when trees
are taken out of the earth in autumn, and allowed
to remain exposed to a dry air all the winter, they
either perish, or are greatly enfeebled.   If, on the
other hand, the soil in which they stand is filled
with moisture, their system is distended with
aqueous matter at a time when it cannot be decomposed or thrown off, and the plant either becomes unnaturally susceptible of the influence of
cold in rigorous climates (112.) or is driven prematurely into growth, when its new parts perish
from the unfavourable state of the air in which
they are then developed.   The most suitable condition of the soil, at the period of vegetable rest,
seems to be that in which no more aqueous matter
is contained than results from the capillary attraction of the earthy particles.

Nevertheless, there are exceptions to this, in the
case of aquatic and marsh plants, whose peculiar
constitution enables them to bear with impunity,
during their winter, an immersion in water ; and
in that of many kinds of bulbs, which, during
their season of rest, are exposed to excessive heat.
The latter plants are, however, constructed in a
peculiar manner ; their roots are annual, and perish
at the same time as the leaves, when the absorbent

organs are all lost, so that the bulb cannot be supposed to require any supply of moisture, inasmuch as it possesses no means of taking it up, even if it existed in the soil. This will be again adverted to in a future chapter.

It is when plants are in a state of growth that an abundant supply of moisture is required in the earth. As soon as young leaves sprout forth, perspiration commences (70.), and a powerful absorption must take place by the roots ; the younger the leaves are, the more rapid their perspiratory action ; their whole epidermis must, at that time, be highly sensible to the stimulating power of light (66.) : but as they grow older their cuticle hardens, the stomates (61.) become the only apertures through which vapour can fly off, and by degrees even these apertures are either choked up, or have a diminished irritability. As a general rule, therefore, we are authorised to conclude that the ground should be abundantly supplied with moisture when plants first begin to grow, and that the quantity should be diminished as the organisation of a plant becomes completed. There are, however, some especial cases which appear to be exceptional, in consequence of the unnatural state in which we require plants to be preserved for our own peculiar purposes. One of the effects of an excessive supply of moisture is, to keep all the newly formed parts of a plant tender and succulent,

and therefore such a constant supply is desirable
when the leaves of plants are to be sent to table,
as in the case of Spinach, Lettuces, and other
oleraceous annuals. Another effect is, to render
all parts naturally disposed to be succulent much
more so than they otherwise would be ; thus we
find market-gardeners deluging their Strawberry
plants with water while the fruit is swelling, in
order to assist in that, to them, important operation. While, however, in this case, the size of
the fruit is increased by a copious supply of water
to the earth, its flavour is, in proportion, diminished ; for, in consequence of the rapidity with
which the strawberry ripens, and, perhaps, the
obstruction of light by its leaves, the excess of
aqueous matter taken into the system cannot be
decomposed, and formed into those products which
give flavour to fruit ; but it must necessarily remain in an unaltered condition.

It is for the reason just given, that the quantity
of water in the soil should be diminished when
succulent fruit is ripening ; we see this happen in
nature, all over the world, and there can be no
doubt of its being of great importance. Not only
is the quality of such fruit impaired by a wet soil,
as has just been shown, but, because of its low
perspiratory power, the fruit will burst from excess
of moisture, as occurs to the plum and grape in
wet seasons. The melon, although an apparent

exception to this rule, is not really so; that fruit
acquires its highest excellence in countries where
its roots are always immersed in water, as in the
floating islands of Cashmere, the irrigated fields
of Persia, and the springy river-beds of India.
But it is to be remembered that the leaves of this
plant have an enormous perspiratory power, arising
partly from their large surface, and partly from
the thinness and consequent permeability of their
tissue, so that they require a greater supply of
fluid than most others; and, in the next place, the
heat and bright light of such countries are capable
of decomposing and altering the fluids of the fruit
with a degree of rapidity and force to which we
can here have no parallel. In this country the
melon does not succeed if its roots are immersed
in water, as I ascertained some years ago, in the
garden of the Horticultural Society, by repeated
experiments. Melons were planted in earth placed
on a tank of water, into which their roots quickly
made their way; they grew in a curvilinear iron
hot-house, and were trained near to the glass,
and consequently were exposed to all the light
and heat that can be obtained in this country. They grew vigorously and produced their
fruit, but it was not of such good quality as it
would have been had the supply of water to the
roots been less copious. Thus, in the tropics, the
quantity of rain that falls in a short time is enor-

mous ; and plants are forced by it into a rapid and
powerful vegetation, which is acted upon by a
light and temperature bright and high in proportion, the result of which is the most perfect organisation of which the plants are susceptible:
but, if the same quantity of water were given to
the same plants at similar periods in this country,
a disorganisation of their tissue would be the result, in consequence of the absence of solar light
in sufficient quantity.

The effect of continuing to make plants grow
in a soil more wet than suits them is well known
to be not only a production of leaves and ill-
formed shoots, instead of flowers and fruit, but,
if the water is in great excess, of a general yellowness of appearance, owing, as some chemists
think, to the destruction, by the water, of a blue
matter which, by its mixture with yellow, forms
the ordinary verdure of vegetation.   If this condition is prolonged, the vegetable tissue enters
into a state of decomposition, and death ensues.
In some cases the joints of the stem separate, in
others the plant rots off at the ground, and all
such results are increased in proportion to the
weakness of light, and the lowness of temperature.
De Candolle considers that the collection of stagnant water about the neck of plants prevents the
free access of the oxygen of the air to the roots ;
but it seems to me that much more mischief is

produced by the coldness of the soil in which water is allowed to accumulate. It seems also probable that the extrication of carburetted hydrogen gas is one cause of the injury sustained by plants whose roots are surrounded by stagnant water; but upon this point we want much more satisfactory evidence than we yet possess.

It is because of the danger of allowing any accumulation of water about the roots of plants that drainage is so very important. In very bibulous soils this contrivance is unnecessary; but in all those which are tenacious, or which from their low situation, do not permit superfluous water to filter away freely, such a precaution is indispensable. No person has ever seen good fruit produced by trees growing in lands imperfectly drained; and all experienced gardeners must be acquainted with cases where wet unproductive borders have been rendered fruitful by contrivances which are only valuable because of their efficiency in regulating the humidity of the soil. Mr. Hiver (*Gard. Mag.*, v. 60.) speaks of the utility of mixing stones in great quantities with the soil, " as they prevent the accumulation of water in very wet weather, and retain sufficient moisture for the purposes of the plant in dry seasons ; " and, when we hear of such precautions as are detailed in the following good account of preparing a Vine border, we only learn how important it is to provide effectually for the removal of superfluous water from around the roots, and how useless a waste of money is that which is expended in forming deep rich beds of earth.

" In preparing a Vine border," says Mr. Griffin, of Woodhall, a successful grower of grapes, " one foot in depth of the mould from the surface is cleared out from the whole space ; a main drain is then sunk parallel to the house, at the extremity of the border, one foot lower than the bottom of the border ; into this, smaller drains are carried dia-gonally from the house across the border. The drains are filled with stone. The cross drains keep the whole bottom quite dry ; but if the subsoil be gravel, chalk, or stone, they will not be necessary. The drainage being complete, the whole bottom is covered with brick, stone, or lime rubbish, about six inches thick, and on this is laid the compost for the vines." (*Hort. Trans.*, iv. 100.)

The practice of placing large quantities of pot-sherds or broken tiles at the bottom of tubs, or pots, or other vessels in which plants are rooted, is only another exemplification of the great necessity of attending to the due humidity of the soil, and to the prevention of stagnant water collecting about the roots ; and the injury committed by worms, upon the roots of plants in pots, is chiefly produced by these creatures reducing the earth to a plastic state, and dragging it among the potsherds so as to stop up the passage between them, and destroy the drainage.

One of the means of guarding the earth against an access on the one hand, and a loss on the other, of too much water, is by paving ground with tiles or stones, and the advantages of this method have been much insisted upon. But it is certain that, in cold summers at least, such a pavement prevents the soil from acquiring the necessary amount of bottom heat; and it is probable that, what with this effect, and the obstruction of a free communication between the atmosphere and the roots of a plant, the practice is disadvantageous rather than the reverse.

More commonly recourse is had to the operation of simple watering, for the purpose of maintaining the earth at a due state of humidity, and to render plants more vigorous than they otherwise would be ; an indispensable operation in hot-houses, but of less moment in the open air. It is, indeed, doubtful whether, in the latter case, it is not often more productive of disadvantage than of real service to plants. When plants are watered naturally, the whole air is saturated with humidity at the same time as the soil is penetrated by the rain; and in this case the aqueous particles mingled with the earth are very gradually introduced into the circulating system : for the moisture of the air prevents a rapid perspiration. Not so when plants in

the open air are artificially watered. This operation
is usually performed in hot dry weather, and must
necessarily be very limited in its effects ; it can have
little if any influence upon the atmosphere : then,
the parched air robs the leaves rapidly of their
moisture, so long as the latter is abundant ; the
roots are suddenly and violently excited, and after
a short time the exciting cause is suddenly with-
drawn, by the momentary supply of water being
cut off by evaporation, and by filtration through
the bibulous substances of which soil usually consists. Then again, the rapid evaporation from the
soil in dry weather has the effect of lowering the
temperature of the earth, and this has been before
shown to be injurious (p. 123.) ; such a lowering,
from such a cause, does not take place when plants
are refreshed by showers, because at that time the
dampness of the air prevents evaporation from the
soil, just as it prevents perspiration from the leaves.
Moreover, in stiff soils, the dashing of water upon
the surface has after a little while the effect of
" puddling " the ground and rendering it impervi-
ous, so that the descent of water to the roots is
impeded, whether it is communicated artificially, or
by the fall of rain. It is, therefore, doubtful whether artificial watering of plants in the open air is
advantageous, unless in particular cases ; and most
assuredly, if it is done at all, it ought to be much
more copious than is usual. It is chiefly in the

case of annual crops that watering artificially is
really important; and with them, if any means of
occasionally deluging ground can be devised, by
means of sluices or otherwise, in the same way as
water meadows, it may be expected to be advantageous. Mildew, which is so often produced by a
dry air acting upon a delicate surface of vegetable
tissue, is completely prevented in annuals by very
abundant watering. The ravages of the Botrytis
effusa, which attacks Spinach; of Acrosporium
monilioides, which is found on the Onion; and
the mildew of the Pea, caused by the ravages of
Erysiphe communis, may all be stopped, or prevented, by abundant watering in dry weather. Mr.
Knight first applied this fact to the securing a late
crop of peas for the table, in the following manner : —

The ground is dug in the usual way, and the
spaces which will be occupied by the future rows
are well soaked with water. The mould upon each
side is then collected, so as to form ridges seven or
eight inches above the previous level of the ground,
and these are well watered ; after which, the seeds
are sowed, in single rows, along the tops of the
ridges. The plants very soon appear above the
soil ; and grow with much vigour, owing to the
great depth of the soil, and abundant moisture.
Water is given rather profusely once in every
week or nine days, even if the weather proves

showery; but, if the ground be thoroughly drenched with water by the autumnal rains, no further trouble is necessary. Under this mode of management, the plants will remain perfectly green and luxuriant till their blossoms and young seed-vessels are destroyed by frost, and their produce will retain its proper flavour, which is always taken away by mildew. (*Hort. Trans.*, ii. 87.)

CHAP. III.

OF ATMOSPHERICAL MOISTURE AND TEMPERATURE. *

THE constituent parts of the atmosphere that surrounds us are either the same in different regions, or the differences, if any, are not appreciable by chemical processes. It is far otherwise, as regards temperature and humidity, which are so intimately connected that they cannot be considered apart from each other.

From what has been already stated (Book I. Chap.

* This subject has already been fully treated by Professor Daniell, in his excellent paper " On Climate with regard to Horticulture," published in the *Transactions of the Horticultural Society*, vol. vi. p. 1. It is impossible for any one to discuss the same topic without profiting largely by this important treatise, which I have very much followed in the present chapter.

IV.), it is apparent that of the vital functions of plants none are more important than those of perspiration and evaporation ; and that, while a certain amount of loss of their fluid particles is necessary, a great excess or diminution of the loss must be injurious. Although the solar rays appear to be the immediate cause of perspiration, which proceeds in proportion to their intensity (71.), yet this action is necessarily modified by the state of the medium, that is, of the atmosphere, which surrounds them ; in proportion to its heat and dryness will their power be augmented, and in proportion to its cold and moisture diminished. The physiological effect of an excessive augmentation of perspiration is to dry up the juices and to destroy the texture of the leaves; on the other hand, an excessive obstruction of that function prevents the decomposition and assimilation of the fluids, and the formation of new organised matter, as well as of the secretions peculiar to a species. A state of the atmosphere, therefore, which is most favourable to the maintenance of the perspiratory action in the most healthy state, is that which it must be the business of a gardener to secure by all the means in his power.

Among the hygrometers intended for measuring the quantity of elastic vapour in the atmosphere, the most convenient for use is that invented by Professor Daniell. In this instrument, the

K

amount of moisture in a given atmosphere is indicated by what is called the *dew-point*; that is to say, by the point of the thermometric scale at which the cold is sufficient to cause a deposition of dew : the amount being calculated by the difference between the natural temperature and an artificial temperature created for the purpose of determining the point at which the elastic vapour of the air is precipitated by cold. "The natural scale of the hygrometer," says Mr. Daniell, "is included between the points of perfect dryness and perfect moisture : the latter, of course, being that state of the atmosphere at which the *dew-point* coincides with the temperature of the air. The intermediate degrees may be ascertained by dividing the elasticity of vapour at the temperature of the dew-point, by the elasticity at the temperature of the air : the quotient will express the proportion of moisture actually existing, to the quantity which would be required for saturation ; for, calling the term of saturation 1·000, as the elasticity of vapour at the temperature of the air is to the elasticity of vapour at the temperature of the dew-point, so is the term of saturation to the actual degree of moisture."

By means of this and similar contrivances*, we

* Other hygrometers have been invented to answer the same end ; but, as Mr. Daniell's is that most eligible in this country, I have thought it more convenient to confine my observations to it.

are at all times able to ascertain exactly the quantity of water that exists in an elastic state in the air. In this country, the changes of moisture are said to extend from 1·000, or saturation, to 389, or even so low as 120, under a south wall, for a short space of time; "a state of dryness which is certainly not surpassed by an African harmattan," but one which produces less disastrous consequences, because it is accompanied by a far lower temperature and a weaker solar radiation. The mean degree of moisture of the air near London has been found by Mr. Thompson to be ·897, on an average of ten years, while the mean temperature is 50·62*: in other parts of the world it is very different; and the amount of those differences, together with the means of imitating them artificially, constitutes one of the most delicate and difficult parts of the gardener's art. All that relates to this subject, however, to be treated usefully, must be considered in a very special way, and in such detail as can only be expected in a separate work upon the subject. An idea of the difference between the atmospherical moisture of London and that of other parts of the world may, however, be collected from the following table showing the amount of rain that falls in a few different countries.

* See the various meteorological journals published by the Horticultural Society, in their *Transactions,* from the year 1826 inclusive.

K 2

Inches per Annum.

LONDON - - -	24·01 Average of 10 years.
St. Petersburg - -	16·
Algiers - - -	27·
Fattehpúr (East Indies)	35·94 Average of 4 years.
Madeira - -	31·
Ságar (East Indies) -	from 31·15 to 64·76
Bahamas - - -	54·99
Calcutta - -	from 59·83 to 81·
Ceylon - - -	84·3
Macao -	from 48·8 to 107·3
Equator - -	96·
Coast of Malabar - -	123·50 Average of 14 years.
Grenada - -	126·
Leogane, St. Domingo -	150·

Bengal, 20 to 22 inches in a single month.
Bombay, 32 inches in 12 days.
Tavoy, 203·5 inches in six months ; as much as
        8·5 in a day (July 31. 1831).

We possess, to a certain extent, the power of modifying the moisture of the air even in the open air, and have almost complete control over that of glazed houses.

It is found by experience that the effect of wind is to increase the dryness of the air, and, consequently, the perspiration of vegetable surfaces. " Evaporation," says Mr. Daniell, "increases in a prodigiously rapid ratio with the velocity of the wind; and any thing which retards the motion of the latter is very efficacious in diminishing the amount of the former. The same surface which, in a calm state of the air, would exhale 100 parts of moisture, would yield 125 in a moderate breeze, and 150 in a high wind." Hence the great importance, in gardens, of walls and screens, which break the wind, and keep the air in repose in their vicinity. The difference between the effect of a given amount of cold upon the blossoms of exposed fruit trees, and those of the same species trained upon walls, is well known; and appears to be owing to this circumstance, much more than to any difference of temperature in the two situations.*

It is to be remarked that the easterly winds are, in this country, both the coldest and the driest. Mr. Daniell tells us that the " moisture of the air flowing from any point between N.E. and S.E. inclusive, is, to that of the air from the other quarters of the compass, in the proportion of ·814 to ·907, upon an average of the whole year;" and Mr. Thompson has found the hygrometer to indicate not uncommonly from 20° to 30° of dryness, during the long prevalence of the north-easterly winds in spring. At the same time, the air is very cold, the effect of which is to cause the sap-vessels

* This has been illustrated by Mr. Howard, in the results of some interesting experiments made by him on the annual amount of evaporation. During three years, in which the evaporating gauge was placed forty-three feet from the ground, the annual average result was 37·85 inches ; during other three years, when the instrument was lower and less exposed, the average was 33·37 inches ; and when the gauge was upon or near the ground, the annual average was only 20·28 inches, or little more than half the amount evaporated in a free and elevated exposure.

of the stem to contract, and refuse to convey their fluid ; so that the blossoms of fruit trees in a north-east wind, while they are robbed of their fluid contents by evaporation, can get no assistance from the roots through the stem, and necessarily perish. I find, however, from Mr. Thompson's observations, that the greatest dryness we experience in this climate is, not when the wind is in the east, but when it is in the south. For example : in nine years, between 1826 and 1834, the four driest days were, in the year 1834, in June, when it was 33° on the 1st, 35° on the 2d, and 31° on the 21st ; on the 1st of June, 1833, it was 30°, and always with a south wind ; and, during the whole of those nine years, there was but one other day on which the dryness was found as high as 30°, namely, on the 10th of April, 1834, with a north-east wind. The duration of dryness, with a south wind, was, however, very short, not exceeding one or at most two days, and was invariably accompanied with great heat and followed by heavy rain ; while the north-easters last for weeks, without rain and with a comparatively low temperature. The following statement by Mr. Thompson puts this in a clear light. There occurred between 1826 and 1834 inclusive, —

Wind North   -   7 days, above 20° of dryness.

    N. East   -   ⌈ 39   do.   do.

    East   114 ⟨ 48   do.   do.

    S. East   -   ⌊ 27   do.   do.

Wind South   -   35 days, above 20° of dryness.
  S. West   -   30 do.   do.
  West   -   35 do.   do.
  N. West   -   22 do.   do.

These facts sufficiently explain the fatal effects of certain winds upon vegetation, the small comparative value in this country of walls with north and east aspects, and the general want of success that attends late spring planting. Here, also, we in part discover an explanation of the utility of shades interposed between the sun and plants newly committed to the earth; they not only cut off the solar rays, but also intercept currents of air, and thus diminish the amount of perspiration by two opposite methods.

The following table, for which I am again indebted to Mr. Thompson, will be found to show that the average degree of dryness, in the middle of the day, throughout the year is, with a

	Degrees of Dryness.		Amount of Moisture.
North wind - - -	6·55°	-	- 816
North-east - - -	7·30	-	- 794
East - - - -	6·20	-	- 825
Average, with wind from the three coldest points	6·68	-	- 811
South wind - - -	4·23	-	- 877
South-west - - -	4·70	-	- 859
West - · - -	6·20	-	- 733
Average, with wind from the three warmest points	5·04	-	- 823

A TABLE

Showing the Temperature, Dryness, and Moisture of the Air, with relation to the Wind, for the Year 1831; calculated by Mr. Robert Thompson, from the Meteorological Observations made in the Garden of the Horticultural Society of London.

1831.	North					North-East					East					South-East				
	Thermometer			Hygrom.		Thermometer			Hygrom.		Thermometer			Hygrom.		Thermometer			Hygrom.	
	Mean Maxima.	Mean Minima.	Media.	Mean Degree of Dryness at Noon.	Mean Degree of Moisture at Noon.	Mean Maxima.	Mean Minima.	Media.	Mean Degree of Dryness at Noon.	Mean Degree of Moisture at Noon.	Mean Maxima.	Mean Minima.	Media.	Mean Degree of Dryness at Noon.	Mean Degree of Moisture at Noon.	Mean Maxima.	Mean Minima.	Media.	Mean Degree of Dryness at Noon.	Mean Degree of Moisture at Noon.
January	34·2	29·0	31·6	3·7	882	37·0	27·1	32·0	3·3	893	43·6	34·5	39·0	0·3	989	44·3	38·3	41·3	0·0	1000
February	43·0	35·0	39·0	6·0	815	44·0	31·0	37·5	12·0	657	43·6	31·3	37·4	0·3	992	-	-	-	-	-
March	-	-	-	-	-	47·0	40·0	43·5	10·6	688	49·6	37·5	43·5	8·0	752	-	-	-	-	-
April	63·5	43·0	53·2	9·5	747	58·8	38·3	48·5	7·5	778	61·7	45·5	53·6	4·2	870	61·6	45·0	53·3	7·6	775
May	73·4	47·0	60·0	10·3	718	67·2	44·0	55·0	11·6	687	64·0	42·5	53·0	15·7	574	61·5	42·0	51·7	8·0	767
June	66·0	49·0	57·5	10·0	721	74·0	41·0	57·5	17·0	572	-	-	-	-	-	-	-	-	-	-
July	-	-	-	-	-	73·0	49·0	61·0	11·0	703	78·5	51·0	64·7	13·2	662	-	-	-	-	-
August	75·8	53·8	64·8	8·0	773	75·5	57·0	65·2	5·5	836	77·5	56·7	67·1	16·7	690	-	-	-	-	-
September	65·2	48·2	56·6	3·0	907	65·0	49·0	57·0	0·0	1000	67·0	48·3	57·6	10·0	723	-	-	-	-	-
October	-	-	-	-	-	-	-	-	-	-	-	-	-	-	-	61·0	45·0	53·0	3·0	904
November	-	-	-	-	-	36·0	27·0	31·5	0·0	1000	58·0	45·0	51·5	0·0	1000	46·0	27·0	36·5	0·0	1000
December	44·0	34·0	39·0	1·0	971	40·0	26·5	33·2	2·5	920	46·0	29·0	37·5	0·0	1000	43·0	29·0	36·0	0·0	1000
Means	58·1	42·3	50·2	6·5	816	55·9	39·0	47·4	7·3	749	58·9	42·1	50·5	6·2	825	52·9	37·7	45·3	3·2	907

1831.	South — Thermometer Mean Maxima	South — Thermometer Mean Minima	South — Thermometer Media	South — Hygrom. Mean Degree of Dryness at Noon	South — Hygrom. Mean Degree of Moisture at Noon	South-West — Thermometer Mean Maxima	South-West — Thermometer Mean Minima	South-West — Thermometer Media	South-West — Hygrom. Mean Degree of Dryness at Noon	South-West — Hygrom. Mean Degree of Moisture at Noon	West — Thermometer Mean Maxima	West — Thermometer Mean Minima	West — Thermometer Media	West — Hygrom. Mean Degree of Dryness at Noon	West — Hygrom. Mean Degree of Moisture at Noon	North-West — Thermometer Mean Maxima	North-West — Thermometer Mean Minima	North-West — Thermometer Media	North-West — Hygrom. Mean Degree of Dryness at Noon	North-West — Hygrom. Mean Degree of Moisture at Noon
January	41·0	30·8	35·9	0·3	982	33·0	25·0	29·0	0·0	1000	41·0	33·6	37·3	0·6	983	—	—	—	—	—
February	46·0	33·0	39·5	1·1	963	59·1	43·8	51·4	3·6	874	48·7	34·0	41·3	6·1	804	50·0	37·0	43·5	0·0	1000
March	54·7	39·0	46·8	2·7	913	54·0	38·6	46·3	5·0	846	55·9	39·0	47·4	8·5	846	60·0	49·0	54·0	3·0	902
April	62·0	44·6	53·3	10·0	711	62·0	48·0	55·0	0·0	1000	57·0	36·5	45·0	—	752	51·0	28·0	39·5	13·0	651
May	67·8	42·1	54·9	7·1	798	74·2	52·5	63·3	12·4	664	—	—	—	—	—	74·0	47·0	60·5	14·0	673
June	—	—	—	—	—	74·0	54·0	64·0	8·3	750	70·6	50·8	60·7	10·8	707	79·0	51·5	65·2	16·0	599
July	—	—	—	—	—	76·0	59·0	67·0	10·0	724	78·0	53·4	65·7	11·9	684	63·5	50·0	60·2	6·0	826
August	77·6	56·0	66·8	9·1	776	69·6	46·0	57·8	5·3	853	67·5	51·6	64·8	12·9	666	—	47·5	55·5	3·0	905
September	70·9	52·4	61·6	6·0	813	63·2	43·6	53·4	4·6	862	61·5	49·5	59·2	8·5	761	50·3	—	—	—	—
October	65·3	51·3	58·3	4·8	885	51·7	38·2	44·9	2·0	938	48·5	39·1	55·5	2·0	939	47·0	28·0	39·1	5·6	938
November	53·6	46·6	50·1	0·6	980	48·5	37·6	43·0	0·9	939	47·6	37·3	43·8	1·6	940	—	42·5	44·7	0·0	1000
December	52·2	42·0	47·1	0·6	980	—	—	—	—	—	—	—	42·9	0·3	986	—	—	—	—	—
Means	59·1	43·7	51·4	4·2	877	60·4	44·2	52·2	4·7	859	59·4	43·2	51·3	6·2	733	54·5	38·0	46·2	6·0	749

As to temperature in the open air, unconnected with atmospherical humidity, there seems to be no means of regulating or modifying it to any considerable extent. In some respects, however, we have even this powerful agent under our control; but, in order to exercise such control, it is necessary to understand correctly the theory of what is called radiation, which cannot be better explained than in the words of Mr. Daniell. "The power of emitting heat in straight lines in every direction, independently of contact, may be regarded as a property common to all matter; but differing in degree in different kinds of matter. Co-existing with it, in the same degrees, may be regarded the power of absorbing heat so emitted from other bodies. Polished metals and the fibres of vegetables may be considered as placed at the two extremities of the scale upon which these properties in different substances may be measured. If a body be so situated that it may receive just as much radiant heat as itself projects, its temperature remains the same; if the surrounding bodies emit heat of greater intensity than the same body, its temperature rises, till the quantity which it receives exactly balances its expenditure, at which point it again becomes stationary; and if the power of radiation be exerted under circumstances which prevent a return, the temperature of the body declines. Thus, if a thermometer be placed in the focus of a concave metallic mirror, and turned towards any clear portion of the sky, at any period of the day, it will fall many degrees below the temperature of another thermometer placed near it, out of the mirror ; the power of radiation is exerted in both thermometers, but to the first all return of radiant heat is cut off, while the other receives as much from the surrounding bodies, as itself projects. This interchange amongst bodies takes place in transparent *media* as well as *in vacuo ;* but in the former case, the effect is modified by the equalising power of the medium.

" Any portion of the surface of the globe which is fully turned towards the sun receives more radiant heat than it projects, and becomes heated ; but when, by the revolution of the axis, this portion is turned from the source of heat, the radiation into space still continues, and, being uncompensated, the temperature declines.   In consequence of the different degrees in which different bodies possess this power of radiation, two contiguous portions of the system of the earth will become of different temperatures ; and, if on a clear night we place a thermometer upon a grass-plat, and another upon a gravel walk or the bare soil, we shall find the temperature of the former many degrees below that of the latter.   The fibrous texture of the grass is favourable to the emission of the heat, but the dense surface of the gravel seems to retain and fix it.   But this unequal effect will only be perceived when the atmosphere is unclouded, and a free passage is open into space ; for even a light mist will arrest the radiant matter in its course, and return as much to the radiating body as it emits. The intervention of more substantial obstacles will of course equally prevent the result, and the balance of temperature will not be disturbed in any sub-stance which is not placed in the clear aspect of the sky.   A portion of a grass-plat under the pro-tection of a tree or hedge, will generally be found, on a clear night, to be eight or ten degrees warmer than surrounding unsheltered parts ; and it is well known to gardeners that less dew and frost are to be found in such situations, than in those which are wholly exposed." (*Hort. Trans.*, vi. 8.)

These laws plainly direct us to the means we are to employ to moderate atmospherical tempera-ture.   A screen, of whatever kind, interposed be-tween the sun and a plant, intercepts the radiant heat of the sun, and returns it into space ; and thus, in addition to the diminution of perspiration by the removal of a part of the stimulus that causes it, actually tends to lower the temperature that sur-rounds the plant.   In like manner, the interposition of a screen, however slight, between a plant and the sky, intercepts the radiant heat of the earth ; and, instead of allowing it to pass off into space, returns it to the ground, the temperature of which is maintained at a higher point than it otherwise would be.   Hence it is that plants growing below the deep projecting eaves of houses, or guarded by a mere coping of thatched hurdles, suffer less in winter than if they were fully exposed to the sky.

It is also obvious from what has been stated, that plants growing upon grass will be exposed to a greater degree of cold in winter than such as grow upon gravel : but it does not therefore follow that hard gravel is, with respect to vegetation, a better coating for the surface of the ground than turf; it has its disadvantages as well as its advantages, and the former probably outweigh the latter. Its superior heating power is its only advantage ; the objections to it are, its dryness in summer, and its comparative impermeability to rain, so that it causes the force of perspiration to be inversely as the absorbing power of the roots.

It is well known that blackened surfaces absorb heat much more than those of any other colour ; and it has been expected that the effect of blackening garden walls, on which fruit trees are trained, would be to accelerate the maturation of the fruit : but, notwithstanding a few cases of apparent advantage, one of which, of the Vine, is mentioned in the *Horticultural Transactions*, vol. iii. p. 330., this has been, in general, found either not to happen at all, or to so small an extent as not to be worth the trouble. It is true, that so long as the wall is but little covered by the branches and leaves of a plant, the absorbent power of the blackened surface is brought into play ; but this effect is lost as soon as the wall becomes covered with foliage. In the early spring, however, before the leaves appear, the flowers are brought rather more forward than would otherwise be the case; and in the autumn the wood certainly becomes more completely ripened, a result of infinite consequence in the northern parts of the country.

It is rather to a judicious choice of soil and situation that the gardener must look for the means of softening the rigour of climate. Wet tenacious soils are found the most difficult to heat or to drain, and they will, therefore, be the most unfavourable to the operations of the gardener ; extremely light sandy soils, on the other hand, part with their moisture so rapidly, and absorb so much heat, that they are equally unfavourable ; and it is the light loamy soils, which are intermediate between the two extremes, that, as is well known, form the best soil for a garden. Situation is, however, of much more consequence than soil, for the latter may be changed or improved, but a bad (that is, cold) situation is incurable. Cold air is heavier than warm air, and, consequently, the stratum of the atmosphere next the soil will be in general colder than those above it. When, there-

fore, a garden is placed upon the level ground of
the bottom of a valley, whatever cold air is formed
upon its surface remains there, and surrounds the
herbage : and, moreover, the cold air that is
formed upon the sides of low hills rolls down into
the valley as quickly as it is formed.   Hence the
fact which to many seems surprising, that what are
called sheltered places are, in spring and autumn,
the coldest.   We all know that the Dahlias, Potatoes, and Kidneybeans of the sheltered gardens
in the valley of the Thames, are killed in the
autumn by frosts whose effects are unfelt on the
low hills of Surrey and Middlesex.   Mr. Daniell
says he has seen a difference of 30°, on the same
night between two thermometers, placed, the one
in a valley, and the other on a gentle eminence,
in favour of the latter.   Hence, he justly observes,
the advantages of placing a garden upon a gentle
slope must be apparent ; " a running stream at its
foot would secure the further benefit of a contiguous surface not liable to refrigeration, and would
prevent any injurious stagnation of the air."

In addition to this, it has been said that, to
obtain the most favourable conditions of climate
in this country, a garden should have a south-
eastern exposure.   This, however, has been recommended, I think, without full consideration.   It is
true that in such an exposure the early sunbeams
will be received ; but, on the other hand, vegetation

there would be exposed to several unfavourable actions. There would be little protection from easterly winds, which, whether south-east or north-east, are the coldest and driest that blow : in the next place, an exposure to the first sun of the morning is very prejudicial to garden productions that have been frozen by the radiation of the night ; it produces a sudden thaw, which, as gardeners well know*, causes the death of plants which, if slowly thawed, would sustain no inconvenience from the low temperature to which they had been exposed. It is probable, as I have elsewhere endeavoured to show, that this singular effect may be accounted for as follows : — "In such cases, it may be supposed that the air, forced into parts not intended to contain it, is expanded violently, and thus increases the disturbance already produced by its expulsion from the proper air cavities ; while, on the other hand, when the thaw is gradual, the air may retreat by degrees from its new situation without producing additional derangement of the tissue. It is also possible that leaves from which their natural air has been expelled by the act of freezing, may, from that circumstance, have their tissue too little protected from the evaporating force of the solar rays, which, we know, produce a specific stimulus of a powerful kind upon those organs." (*Hort. Trans.*, n. s., ii. 305.)

* See Hort. Trans., iii. 43.

In our glazed houses, we have full control over the state of the atmosphere, as regards both its moisture and temperature, by means familiar to every gardener ; but the manner of applying those means, and the causes that oppose their action, deserve to be the subject of enquiry.

It will have been seen, from what has been already stated upon that subject, that in general, in warm countries, the air is occasionally at least, if not permanently, filled with vapour to a much greater extent than in northern latitudes*, and, as in our glazed houses we cultivate exclusively the natives of warm countries, it is also obvious that, as a general rule, the air of such houses requires, at certain periods, to be damper than that of the external air. Those periods are when vegetation is most active. On the other hand, countries nearer the equator are subject to seasons of dryness, the continuance of which is often much greater than any thing we know of here in the open air, and consequently artificial means must also be adopted to bring about, in glazed houses, that state of things at particular periods ; namely, those of the repose of plants. These facts afford abundant proof

* " Captain Sabine, in his meteorological researches between the tropics, rarely found at the hottest period of the day so great a difference as 10° between the temperature of the air and the dew-point ; making the degree of saturation about 730, but most frequently 5° or ·850 ; and the mean saturation of the air could not have exceeded ·910." (*Daniell*)

of the necessity of regulating the moisture of the atmosphere with due precision.

By throwing water upon the pavement of glass houses, by means of open tanks of water, by reservoirs placed upon them, by syringing, and by other contrivances*, the quantity of water in the air may be easily increased, even up to the state of saturation. But there are some circumstances, easily overlooked, which interfere very seriously with this power, and which, it may be conceived, may reduce it very much below the expectations of the cultivator.

The most unsuspected of these is the destruction of aqueous vapour by the hot, dry, absorbent surface of flues. The advantages derived from hot-water pipes, or steam pipes, over brick flues, are so well known, as not to require any evidence to prove the fact. Gardeners explain the difference in the action of the two, by saying that the dry heat produced by hot-water pipes is *sweeter* than that given off by flues; which is not a very intelligible expression. The fact is, that the air of houses heated by flues is, under equal circumstances, much drier than that where hot-water pipes are employed; because the soft burnt clay of the brick flues robs the air of

---

* A discharge of steam into a glazed house has occasionally been employed; but the method requires much attention on the part of the operator, and seems inferior to other contrivances.

its moisture, while the unabsorbent surface of iron pipes abstracts nothing.

Another source of dryness is the coldness of the glass roof, especially in cold weather, when its temperature is lowered by the external air, in consequence of which the moisture of the artificial atmosphere is precipitated upon the inside of the glass, whence it runs down in the form of "drip." Mr. Daniell observes that the glass of a hot-house, at night, cannot exceed the mean of the external and internal air ; and, taking them at 80° and 40°, 20 degrees of dryness are kept up in the interior, or a degree of saturation not exceeding ·528. To this, in a clear night, we may add at least 6° for the effects of radiation, to which the glass is particularly exposed, which will reduce the saturation to ·424 ; and this is a degree of drought which must be very prejudicial. It will be allowed that this is not an extreme case, and much more favourable than must frequently occur during the winter season. Some idea, he adds, may be formed of the prodigiously increased drain upon the functions of a plant, arising from an increase of dryness in the air, from the following consideration : — If we suppose the amount of its perspiration, in a given time, to be 57 grains, the temperature of the air being 75° and the dew-point 70°, or the saturation of the air being ·849, the amount would be increased to 120 grains in the same time, if the dew-point were to remain stationary, and the temperature were to rise to 80°; or, in other words, if the saturation of the air were to fall to 726. (*Hort. Trans.*, vi. 20.) It is well known that the effect of maintaining a very high temperature in hot-houses at night, during winter, is frequently to cause the leaves to wither and turn brown, as if scorched or burnt; and this is apparently owing to the dryness of the air, in consequence of the above causes.

It is evident that the mode of preventing this drying of the air by the cold surface of a glass roof will be, either by raising the temperature of the glass, which can only be effected by drawing a covering of some kind over our houses at night, so as to intercept radiation, or by double glass sashes; or else by keeping the temperature of the air of the house as low as possible, consistently with the safety of the plants, and so diminishing the difference between the temperature of the external and internal air.

A bad system of ventilation is another cause of the loss of vapour in the atmosphere of· glazed houses, to which reference will be made in the succeeding chapter.

It is, in all appearance, to the attention that, since the appearance of Mr. Daniell's paper, in 1824, upon this subject, has been paid to the atmospherical moisture of glazed houses, that the great superiority of modern gardeners over those of the last generation is mainly to be ascribed ; there are, however, traces of the practice at a much earlier period, although, from not understanding the theory of the practice, no general improvement took place. In the year 1816, an account was laid before the Horticultural Society of a very successful mode of forcing grapes and nectarines, as practised by Mr. French, an Essex farmer, with very rude materials, and under unfavourable circumstances. It is not a little remarkable, that, although Mr. French himself correctly referred his success to the skilful management of the atmospherical moisture of his forcing-houses, the subject was so little understood at that time, that the author of the account not only shrank from adopting the opinion, but evidently, from the manner in which he speaks of Mr. French's explanation, had no idea of its justness. The method itself is sufficiently remarkable to deserve being extracted.

" About the beginning of March, Mr. French commences his forcing, by introducing a quantity of new long dung, taken from under the cow-cribs in his straw yard ; being principally, if not entirely, cow dung ; which is laid upon the floor of his house, extending entirely from end to end, and in width about six or seven feet, leaving only a pathway between it and the back wall of the house. The dung being all new at the beginning, a profuse steam arises with the first heat, which, in this stage of the process, is found to be beneficial in destroying the ova of insects, as well as trans-fusing a wholesome moisture over the yet leafless branches ; but which would prove injurious, if permitted to rise in so great a quantity when the leaves have pushed forth.   In a few days, the vio-lence of the steam abates as the buds open, and in the course of a fortnight the heat begins to diminish ; it then becomes necessary to carry in a small addition of fresh dung, laying it in the bottom, and covering it over with the old dung fresh forked up : this produces a renovated heat, and a moderate exhalation of moist vapour.   In this manner the heat is kept up throughout the season, the fresh supply of dung being constantly laid at the bottom in order to smother the steam, or rather to moderate the quantity of exhalation ; for it must always be remembered, that Mr. French attaches great virtue to the supply of a reasonable portion of the vapour.   The quantity of new dung to be introduced at each turning must be regu-lated by the greater or smaller degree of heat that is found in the house, as the season or other cir-cumstances appear to require it.   The temperature kept up is pretty regular, being from 65 to 70 degrees." (*Hort. Trans.*, i. 245.)

In this case, which attracted much attention at the time, it is evident that the success of the practice arose principally out of two circumstances: firstly, the moisture of the atmosphere was skilfully maintained in due proportion to the temperature; and, secondly, a suitable amount of bottom heat was secured. This is, as will be elsewhere re-marked, the principal cause of the advantages found to attend the Dutch mode of forcing. The reporter upon Mr. French's practice speaks with surprise of the rudeness of the roof of his forcing-houses, and of the numerous openings into the air through the laps of the glass and the joints of the sashes; but these were points of no importance under the mode of management adopted.

The impossibility of preserving any plants, ex-cept succulents, in a healthy state, for any long period, in a sitting-room, is evidently owing to the impracticability of maintaining the atmosphere of such a situation in a state of sufficient dampness.

An excess of dampness is indispensable to plants in a state of rapid growth, partly because it prevents the action of perspiration becoming too violent, and partly because under such circum-stances a considerable quantity of aqueous food is absorbed from the atmosphere, in addition to that obtained by the roots.

But it is essential to observe that, when not in a state of rapid growth, a large amount of moisture in the air will be prejudicial rather than advan-tageous to a plant; if the temperature is at the

same time high, excitability will remain in a
state of continued action, and that rest which is
necessary (113.) will be withheld, the result of
which will be an eventual destruction of the vital
energies. But, on the other hand, if the temperature is kept low while the amount of atmospherical moisture is considerable, the latter is absorbed,
without its being possible for the plant to decompose it; the system then becomes, in the younger
and more absorbent parts, distended with water,
and decomposition takes place, followed by the appearance of a crop of microscopical fungi; in short,
that appearance presents itself which is technically
called " damping off."

The skilful balancing of the temperature and
moisture of the air, in cultivating different kinds
of plants, and the just adaptation of them to the
various seasons of growth, constitute the most
complicated and difficult part of a gardener's art.
There is some danger in laying down any general
rules with respect to this subject, so much depends
upon the peculiar habits of species, of which the
modifications are endless. It may, however, I
think, be safely stated, that the following rules
deserve especial attention : —

1. Most moisture in the air is demanded by
plants when they first begin to grow, and least when
their periodical growth is completed.

2. The quantity of atmospheric moisture required

by plants is, *cæteris paribus*, in inverse proportion
to the distance from the equator of the countries
which they naturally inhabit.

3. Plants with annual stems require more than
those with ligneous stems.

4. The amount of moisture in the air most suit
able to plants at rest is in inverse proportion to the
quantity of aqueous matter they at that time contain. (Hence the dryness of the air required by
succulent plants when at rest.)

CHAP. IV.

OF VENTILATION.

By far the larger number of gardeners attach great
importance to preserving the power of ventilating
their houses abundantly, without perhaps sufficiently considering the nature of the plants they
have to manage; and, as has been justly enough
said, by supposing that plants require to be treated
like man himself, they consult their own feelings
rather than the principles of vegetable growth.
There can be no doubt, however, that the effect
of excessive ventilation is more frequently injurious
than advantageous; and that many houses, particularly hothouses, would be more skilfully managed,

if the power of ventilation possessed by the gardener were much diminished.

Animals require a continual renovation of the air that surrounds them, because they speedily render it impure by the carbonic acid given off, and the oxygen abstracted by animal respiration. But the reverse is what happens to plants; they exhale oxygen during the day, and inhale the carbonic acid of the atmosphere, thus depriving the latter of that which would render it unfit for the sustenance of the higher orders of the animal kingdom; and, considering the manner in which glass houses of all kinds are constructed, the buoyancy of the air in all heated houses would enable it to escape in sufficient quantity to renew itself as quickly as can be necessary for the maintenance of the healthy action of the organs of vegetable respiration. It, therefore, is improbable that the ventilation of houses in which plants grow is necessary to them, so far as respiration is concerned. Indeed, Mr. Ward has proved that many plants will grow better in confined air, than in that which is often changed. By placing various kinds of plants in cases, made, not indeed air-tight, for that is impossible with such means as can be applied to the construction of a glass house, but so as to exclude as much as possible the admission of the external air, supplying them with a due quantity of water, and exposing them fully to light, he has shown the possibility of cultivating them without ventilation, with much more success than usually attends ordinary glass-house management.

In forcing-houses, in particular, it will be evident from what is about to follow, that ventilation, under ordinary circumstances, in the early spring, must be productive of injury rather than benefit. Many gardeners now admit air very sparingly to their vineries during the time that the leaves are tender, and the fruit unformed. Some excellent stoves have no provision at all for ventilation; and we have the direct testimony of Mr. Knight as to the disadvantage of the practice in many cases to which it has been commonly applied.

It may be objected, says this great horticulturist, that plants do not thrive, and that the skins of grapes are thick, and other fruits without flavour, in crowded forcing-houses : but in these it is probably light, rather than a more rapid change of air, that is wanting ; for, in a forcing-house which I have long devoted almost exclusively to experi ments, I employ very little fire heat, and never give air till my grapes are nearly ripe, in the hottest and brightest weather, further than is just necessary to prevent the leaves being destroyed by excess of heat. Yet this mode of treatment does not at all lessen the flavour of the fruit, nor render the skins of the grapes thick ; on the contrary, their skins are always most remarkably thin, and very similar to those of grapes which have ripened in the open air. (*Hort. Trans.*, ii. 225.)

While, however, the *natural* atmosphere cannot be supposed to require changing in order to adapt it to the respiration of plants, it is to be borne in mind that the air of houses artificially heated may be rendered impure by the means employed to produce heat. Sulphurous acid gas escapes from brick flues, ammoniacal vapour from fermenting manure, and there may be many unsuspected sources of the introduction of vaporous impurities; an inconceivably minute quantity of which is enough to deteriorate the air, so far as vegetation is concerned. Drs. Turner and Christison found that $\frac{1}{10000}$ of sulphurous acid gas destroyed leaves in forty-eight hours; and similar effects were obtained from hydrochloric or muriatic acid gas, chlorine, ammonia, and other agents, the presence of which was perfectly undiscoverable by the smell. We also know that the destructive properties of air poisoned by corrosive sublimate, perhaps by its being dissolved in the vapour of a hothouse, are not at all appreciable by the senses.

Ventilation is necessary, then, not to enable plants to exercise their respiratory functions, provided the atmosphere is unmixed with accidental impurities; but to carry off noxious vapours generated in the artificial atmosphere of a glazed house, and to produce dryness, or cold, or both.

When the external air is admitted into a glazed house containing a moist atmosphere, it, under ordinary circumstances, is much colder than that with which it mixes; the heated damp air rushes out at the upper ventilators, and the drier cold air takes its place; the latter rapidly abstracts from the plants and the earth, or the vessels in which they grow, a part of their moisture, and thus gives a sudden shock to their constitution, which cannot fail to be injurious. This abstraction of moisture is in proportion to the rapidity of the motion of the air. But it is not merely dryness that is thus produced, or such a lowering of temperature as the thermometer suspended in the interior of the house may indicate; the rapid evaporation that takes place upon the admission of dry air produces a degree of cold upon the surface of leaves, and of the porous earthen pots in which plants grow, of which our instruments give no indication. To counteract these mischievous effects many contrivances have been proposed, in order to insure the introduction of fresh air warm and loaded with moisture, such as compelling the fresh air to enter a house after passing through pipes moderately heated, or over hot-water pipes surrounded by a damp atmosphere, and so on, the advantages of which, of course, depend upon the objects to be attained.

If ventilation is merely employed for the purpose of purifying the air, as is often the case in hothouses and in dung pits, it should be effected by the introduction of fresh air damp and heated.

If it is only for the purpose of lowering the temperature, as in greenhouses, or in the midst of summer, the external air may be admitted without any precautions.

But it is very commonly required in the winter, for the purpose of drying the air in houses kept at that season at a low temperature; such, for instance, as those built for the protection of Heaths, and many other Cape and New Holland plants: in these cases it should be brought into the house as near the temperature of the house as possible, but on no account loaded with moisture. One of the principal reasons for drying the air of such houses is, to prevent the growth of parasitical fungi, which, in the form of mouldiness, constitute what gardeners technically call "damp." These productions flourish in damp air at a low temperature, but will not exist either in dry cold air or in hot damp air. If the air of cool greenhouses is allowed to become damp, the fungi immediately spring up on the surface of any decayed leaves, or other matter which may be present, when they spread rapidly to the young and tender parts of living plants ; and when this happens they consume the juices, choke the respiratory organs, and speedily destroy the object they attack.

Ventilation is also required in the winter in such places as dung pits or frames, especially those in which salad, cucumbers, and similar plants are grown. In those cases the object is to dry the air, in order that the plants may not absorb more aqueous particles than they can decompose and assimilate. Although plants of this kind will bear a high degree of atmospherical moisture in summer, when the days are long and the sun bright, and when, consequently (66, 67.), all their digestive energies are in full activity, yet they are by no means able to endure the same amount in the short dark days of winter, when, from the want of light, their powers of decomposition or digestion are comparatively feeble. Hence, no doubt, the advantage of growing winter cucumbers in forcing-houses, instead of dung frames.

One of the causes of success in the Dutch method of winter forcing is, undoubtedly, their avoiding the necessity of winter ventilation, by intercepting the excessive vapour that rises from the soil, and which would otherwise mix with the air. For this purpose they interpose screens of oiled paper between the earth and the air of their houses, and in their pits for vegetables they cover the surface of the ground with the same oiled paper, by which means vapour is effectually intercepted, and the air preserved from excessive moisture.

In forcing-houses, ventilation is thought to be

required at the time when the fruit is ripening, for
the purpose of increasing the perspiration of the
plants, and, consequently, of assisting in the elaboration of the secretions to which fruit owes its
flavour : but, even for this, its utility is, perhaps,
overrated, and it is quite certain that it may be easily
carried to excess; for, if it is so powerful as to injure
the leaves by over-drying the air, an effect the
reverse of what was intended will be produced ;
that is to say, the quality of the fruit will be deteriorated (64. 75.). Upon this subject Mr. Knight
has made the following observations : — " A less
humid atmosphere is more advantageous to fruits
of all kinds, when the period of their maturity approaches, than in the earlier stages of their growth;
and such an increase of ventilation, at this period,
as will give the requisite degree of dryness to the
air within the house, is highly beneficial, provided
it be not increased to such an extent as to reduce
the temperature of the house much below the
degree in which the fruit had previously grown,
and thus retard its progress to maturity.   The
good effect of opening a peach-house, by taking
off the lights of its roof during the period of the
last swelling of the fruit, appears to have led
many gardeners to overrate greatly the beneficial
influence of a free current of air upon ripening
fruits ; for I have never found ventilation to give
the proper flavour or colour to a peach, unless that

fruit was, at the same time, exposed to the sun without the intervention of glass ; and the most excellent peaches I have ever been able to raise were obtained under circumstances where change of air was as much as possible prevented, consistently with the admission of light (without glass), to a single tree." (*Hort. Trans.*, ii. 227.)

It is not improbable that one of the advantages of ventilation depends upon a cause but little adverted to, but which certainly requires to be well considered.

It was an opinion of Mr. Knight, that the motion given to plants by wind is beneficial to them by enabling their fluids to circulate more freely than they otherwise would do; and in a paper printed in the *Philosophical Transactions* for 1803, p. 277., he adduces, in support of his opinion, many experiments and observations ; of which the following is sufficiently striking : —

" The effect of motion on the circulation of the sap, and the consequent formation of wood, I was best able to ascertain by the following expedient. Early in the spring of 1801, I selected a number of young seedling Apple trees, whose stems were about an inch in diameter, and whose height between the roots and first branches was between six and seven feet. These trees stood about eight feet from each other ; and, of course, a free passage for the wind to act on each tree was afforded. By

M

means of stakes and bandages of hay, not so tightly bound as to impede the progress of any fluid within the trees, I nearly deprived the roots and lower parts of the stems of several trees of all motion, to the height of three feet from the ground, leaving the upper part of the stems and branches in their natural state. In the succeeding summer, much new wood accumulated in the parts which were kept in motion by the wind; but the lower parts of the stems and roots increased very little in size. Removing the bandages from one of these trees in the following winter, I fixed a stake in the ground, about ten feet distant from the tree, on the east side of it; and I attached the tree to the stake at the height of six feet, by means of a slender pole, about twelve feet long; thus leaving the tree at liberty to move towards the north and south, or, more properly, in the segment of a circle, of which the pole formed a radius; but in no other direction. Thus circumstanced, the diameter of the tree from north to south in that part of its stem which was most exercised by the wind exceeded that in the opposite direction, in the following autumn, in the proportion of thirteen to eleven."

Now, if the effect of motion is to increase the quantity of wood in a plant, it is evident that ventilation, which causes motion, must tend to produce a healthy action in the plants exposed to it;

and such a state must also be favourable to the developement of all those secretions upon which the organisation of flowers, the setting of fruit, and the elaboration of colour, odour, flavour, &c., so much depend. Some suggestions by Mr. Knight, as to the manner in which this result can be artificially produced, will be found in the *Hort. Trans.*, vol. iv. p. 2. and 3.; but the subject has as yet attracted little attention. (See also *Hort. Trans.*, new series, i. 34.)

CHAP. V.

OF SEED-SOWING.

WHEN a seed is committed to the earth, it undergoes certain chemical changes (14.) before it can develope new parts and grow. These changes are brought about by heat and water, assisted by the absence of light. In many seeds the vital principle is so strong, that to scatter them upon the soil, and to cover them slightly with earth, are sufficient to insure their speedy germination; but in others the power of growth will only manifest itself under very favourable conditions : it is, therefore, necessary to consider well upon what the circumstances most suitable to germination depend.

Moisture is necessary, but not an unlimited

quantity.   If a seed is thrown into water and exposed to a proper temperature, the act of germination will take place : but, unless the plant is an
aquatic, it will speedily perish ; no doubt because
its powers of respiration are impeded, and it is
unable to decompose the water it absorbs, which
collects in its cavities and becomes putrid.   There
must, therefore, be some amount of water, which
to the dormant as well as the vegetating plant
is naturally more suitable than any other ; and
experience shows that quantity to be just so much
as the particles of earth can retain around and
among them by the mere force of attraction.   To
this is to be ascribed the advantage derived from
those mixtures of peat, loam, and sand, which
gardeners prefer for their seedlings ; the peat and
sand together keep asunder the particles of loam
which would otherwise adhere and prevent the
percolation of water ; the loam retains moisture
with force enough to prevent its passing off too
quickly through the wide interstices of sand and
peat. If, during the delicate action of germination,
the changes that the seed undergoes take place
without interruption, the young plant makes its
appearance in a healthy state ; but, if by irregular
variations of heat, light, and moisture, the progress of germination is sometimes accelerated and
sometimes stopped, the fragile machinery upon
which vitality depends may become so much de-

ranged as to be no longer able to perform its actions, and the seed will die. It is for the purpose of securing uniformity in these respects, that we employ, in delicate cases, the steady heat of a gentle hotbed, shaded; and, in all cases whatever, the assistance of a coating of earth scattered over the seed.

Under what depth of earth seed should be buried must always be judged of by the experience of a gardener: but it should be obvious that minute seeds, whose powers of growth must be feeble in proportion to their size, will bear only a very slight covering; while others, of a larger size and more vigour, will be capable, when their vital powers are once put in action, of upheaving considerable weights of soil. As, however, the extent of this power is usually uncertain, the judicious gardener will take care to employ, for a covering, no more earth than is really necessary to preserve around his seeds the requisite degree of darkness and moisture.* Hence the common practices of sow-

* It may, perhaps, be as well to notice, in this place, an erroneous opinion, not uncommonly entertained, that seeds must be "well" buried in order that the young plants, when produced, may have " sufficient hold of the ground." The fact is, that a seed, when it begins to grow, plunges its root downwards and throws its stem upwards from a common point, which is the seed itself; and, consequently, all the space that intervenes between the surface of the soil and the seed is occupied by the base of the stem, and not by roots.

M 3

ing small seeds upon the surface of the soil, and covering them with a coating of moss, which may be removed when the young seedlings are found to have established themselves. In other cases very minute seeds are mixed with sand before they are sown.

The latter practice is not, however, merely for the sake of covering the seed with the smallest possible quantity of soil, but has for its object the separation of seeds to such a distance, that when they germinate they may not choke up each other. If seedlings, like other plants, are placed so near together that they either exhaust the soil of its organisable matter, or overshadow each other so as to hinder the requisite quantity of light, some will die in order that the remainder may live; and this, in the case of rare seeds, should, of course, be guarded against very carefully.

With regard to the temperature to which a seed should be subjected, in order to secure its germination, this, undoubtedly, varies with different species, and depends upon their peculiar habits, and the temperature of the climate of which they are native. So far as general rules can be given upon such a subject, it may be stated that the temperature of the earth most favourable for germination is 50° to 55° for the seeds of cold coun- tries, 60° to 65° for those of "greenhouse plants," and 70° to 80° for those of the torrid zone. No seed, however, has been known to refuse to germinate in the last-mentioned temperature, although those to which such a heat is necessary will not, in general, grow in a healthy manner in a lower temperature. We have no exact experiments upon this subject, except in a few cases recorded by Messrs. Edwards and Colin, by whom there is a very valuable set of observations upon the temperatures borne by certain agricultural seeds (*Annales des Sciences*, new series, vol. v. p. 5.), the result of which may be thus stated : —

At 44·6°, Wheat, Barley, and Rye could germinate.

> 95°, *in water*, for three days, ⅔ of the Wheat and Rye, and all the Barley, were killed.
>
> 104°, *in sand and earth*, the same seeds sustained the temperature for a considerable time, without inconvenience.
>
> 113°, under the same circumstances, most of them perished.
> 122°,   ditto   ditto   all perished.

But it was found that, for short periods of time, a much higher temperature could be borne.

> At 143·6°, *in vapour*, Wheat, Barley, Kidneybeans, and Flax retained their vitality for a quarter of an hour; but in 27½ minutes, the three last died at a temperature of 125·6°.
>
> 167°, *in vapour*, they all perished.
> 167°, *in dry air*, they sustained no injury.

It will be presently seen that some seeds will bear a much higher temperature.

The foregoing observations apply to seeds in a

perfect state of health; when they have become sickly or feeble, from age or other causes, some precautions become necessary, to which, under other circumstances, no attention requires to be paid.

When the vital energies of a seed are diminished, it does not lose its power of absorbing water, but it is less capable of decomposing it (14.). The consequence of this is, that the free water introduced into the system collects in the cavities of the seed, and produces putrefaction; the sign of which is the rotting of seeds in the ground. The remedy for this is to present water to the seed in such small quantities at a time, and so gradually, that no more is absorbed than the languid powers of the seed can assimilate; and to increase the quantity only as the dormant powers of vegetation are aroused. One of the best means of doing this is, to sow seeds in warm soil tolerably dry; to trust for some time to the moisture that exists in such earth, and in the atmosphere, for the supply required for germination; and only to administer water when the signs of germination have become visible; even then the supply should be extremely small. If this is attended to, carbonic acid is very slowly formed and liberated, the chemical quality of the contents of the seed is thus insensibly altered, each act of respiration may be said to invigorate it, and by degrees it will be brought to a condition favourable to the assimilation of food in larger quantities. Mr. Knight used to say that these effects were produced in no way so well as by enclosing seeds between two pieces of loamy turf, cut smooth, and applied to each other by the underground sides ; such a method is however, scarcely applicable to any except seeds of considerable size.

Other expedients have occasionally been had recourse to successfully. Where seeds are enclosed in a very hard dry shell, it is usually necessary to file it thin, so as to permit the embryo to burst through its integuments when it has begun to swell. Under natural circumstances, indeed, no such operation is practised : but it is to be remembered that such seeds will have fallen to the ground as soon as ripe, and before their shell acquired the bony hardness that we find after having become dry.

Sometimes it has been found useful to immerse seeds in tepid water until signs of germination manifest themselves, and then to transfer them to earth: but this process cannot be applied with advantage to seeds in an unhealthy state ; and it is only of use to healthy seeds, by accelerating the time of growth, a practice which may, in outdoor crops, be sometimes desirable, when applied to seeds which, like the Beet, the Carrot, or the Parsnep, will, in dry seasons, lie so long in the ground without germinating, that they become a prey to birds or other animals.

Of late years, the singular practice has been introduced of *boiling* seeds, to promote germination. This was, I believe, first recommended by Mr. Bowie, who stated, in the *Gardener's Magazine*, vol. viii. p. 5. (1832), that " he found the seeds of nearly all leguminous plants germinate more readily by having water heated to 200°, or even to the boiling point of Fahrenheit's scale, poured over them, leaving them to steep and the water to cool for twenty-four hours." Subsequently, the practice has been adopted by other persons with perfect success; and, some years ago, seedlings of Acacia lophantha were exhibited before the Horticultural Society by the late Mr. Thomas Cary Palmer, which had sprung from seeds boiled for as much as five minutes. I am also acquainted with other cases, one of the more remarkable of which was the germination of the seeds of the Raspberry, picked from a jar of jam, and which must therefore have been exposed to the temperature of 230°, the boiling point of syrup. It is difficult to understand in what way so violent an action can be beneficial to any thing possessing vitality; the fact, however, is certain. As such instances of success are confined to seeds with hard shells, it is possible that the heated fluid may act in part mechanically by cracking the shell, in part as a solvent of the matters enclosed in the seed, and in part as a stimulant.

Mr. Lymburn, nurseryman at Kilmarnock, has lately called attention to the effect produced upon germinating seeds by alkaline substances. He states that experiments made by Mr. Charles Maltuen, and narrated in Brewster's *Journal of Science*, having shown that the negative or alkaline pole of a galvanic battery caused seeds to germinate in much less time than the positive or acid pole, he was induced to observe the effects on seeds of acetic, nitric, and sulphuric acids, and also of water rendered alkaline by potash and ammonia. " In the alkaline, the seeds vegetated in thirty hours, and were well developed in forty ; while in the acetic and sulphuric they took seven days ; and, even after a month, they had not begun to grow in the acetic." This experiment led to others upon lime ; " a very easily procured alkali, and which he inferred to be more efficient than any other from the well known affinity of quick or newly slacked lime for carbonic acid. Lime, as taken from the quarry, consists of carbonate of lime, or lime united to carbonic acid ; but, in the act of burning, the carbonic acid is driven off ; and hence the great affinity of newly slacked lime for carbonic acid. He depended, therefore, upon this affinity, to extract the carbon from the starch, assisted by moisture ; " (*Gard. Mag.*, xiv. 74.)

and it is stated that the results were exceedingly striking. Old Spruce Fir seed, which would scarcely germinate at two years old, produced a fine healthy crop when three years old, having been first damped and then mixed with newly slacked lime; and, under the same treatment, an average crop of healthy plants was obtained when the seed was four years old. Unfortunately, the manner in which the original experiments upon acids and alkalies were conducted is not explained (it is to be presumed that the water employed was only *acidulated* with the acids spoken of), and I am not aware of the experiments having been repeated.

The last method of promoting germination, to which it is necessary to advert, is the mixing seeds with agents that have the power of liberating oxygen. It has been shown (14.) that a seed cannot germinate until the carbon with which it is loaded is to a considerable extent removed; the removal of this principle is effected by converting it into carbonic acid, for which purpose a large supply of oxygen is required. Under ordinary circumstances, the oxygen is furnished by the decomposition of water by the vital forces of the seed; but, when those forces are languid, it has been proposed to supply oxygen by some other means. Humboldt employed a dilute solution of chlorine, which has a powerful tendency to decompose water and set oxygen at liberty, and, it is said, with great success. Oxalic acid has also been used for the same purpose.

Mr. Otto, of Berlin, states that he employs oxalic acid to make old seeds germinate. The seeds are put into a bottle filled with oxalic acid, and remain there till the germination is observable, which generally takes place in from twenty-four to forty-eight hours, when the seeds are taken out, and sown in the usual manner. Another way is to wet a woollen cloth with oxalic acid, on which the seeds are put, and it is then folded up and kept in a stove; by this method small and hard seeds will germinate equally as well as in the bottle. Also very small seeds are sown in pots and placed in a hotbed; and oxalic acid, much diluted, is applied twice or thrice a day till they begin to grow. Particular care must be taken to remove the seeds out of the acid as soon as the least vegetation is observable. Mr. Otto found that by this means seeds which were from twenty to forty years old grew, while the same sort, sown in the usual manner, did not grow at all (*Gard. Mag.*, viii. 196.) : and it is asserted by Dr. Hamilton (*Ib.*, x. 368. 453.) and others, that they have found decided advantages from the employment of this substance. Theoretically it would seem that the effects described ought to be produced, but general experience does not confirm them; and it may be conceived that the rapid abstraction of carbon, by the presence of

an unnaturally large quantity of oxygen, may produce effects as·injurious to the health of the seed,
as its too slow destruction in consequence of the
languor of the vital principle.

The length of time that some seeds will lie in
the ground, under circumstances favourable to
germination, without growing, is very remarkable,
and inexplicable upon any known principle. If the
Hawthorn be sown immediately after the seeds are
ripe, a part will appear as plants the next spring; a
larger number the second year; and stragglers,
sometimes in considerable numbers, even in the
third and fourth seasons. Seeds of the genera Ribes,
Berberis, and Pæonia have a similar habit. M. Savi is
related by De Candolle to have had, for more than
ten years, a crop of Tobacco from one original sowing; the young plants having been destroyed yearly,
without being allowed to form their seed. This
matter does not, perhaps, concern the theory of
horticulture, for theory is incapable of explaining
it; but it is a fact that it is useful to know, because
it may prevent still living seeds from being thrown
away, under the idea that, as they did not grow the
first year, they will never grow at all.

CHAP. VI.

OF SEED-SAVING.

THE maturation of the seed, being a vital action indispensable to the perpetuation of a species, is, in wild plants, guarded from interruption by so many wise precautions, that no artificial assistance is required in the process; but in gardens, where plants are often enfeebled by domestication, or exposed to conditions very different from those to which they are subject in their natural state, the seed often refuses to ripen, or even to commence the formation of an embryo. In such cases, the skill of gardeners must aid the workings of nature, and art must effect that which the failing powers of a plant are unable to bring about of themselves.

Sterility is a common malady of cultivated plants; the finer varieties of fruit, and all double and highly cultivated flowers, being more frequently barren than fertile. This arises from several different causes.

The most common cause of sterility is an unnatural developement of some organ in the vicinity of the seed, which attracts to itself the organisable matter that would otherwise be applicable to the support of the seed. Of this the Pear, the Pineapple, and the Plantain are illustrative instances.

The more delicate varieties of Pear, such as the Gansel's Bergamot and the Chaumontelle, have rarely any seeds ; of Pine-apple, none, except the Enville now and then, have seeds, and that variety, though a large one, is of little value for its delicacy, and probably approaches nearly to the wild state of the plant ; of Plantains, few, except the wild and crabbed sorts, are seedful. The remedy for this appears to be, the withholding from such plants all the sources from which their succulence can be encouraged. If, in consequence of any predisposition to form succulent tissue (on which the excellence of fruit much depends), the organisable matter of the plant be once diverted from feeding the seed to those parts in which the succulence exists, it will continue, by the action of endosmose, to be attracted thither more powerfully than to any other part, and the effect of this will be the starvation of the seed : but a scanty supply of food, an unhealthy condition of the plant itself, or withholding the usual quantity of water, will all check the tendency to luxuriance, and therefore will favour the developement of the seed, whose feeble attracting force is, in that case, not so likely to be overcome by the accumulation of attracting power in the neighbouring parts. Thus we see that Pine-apples are more frequently seedful under the bad cultivation of the Continent, than in the highly kept and skilfully managed pineries of England.

Abstraction of branches, in the neighbourhood of fruit, has also been occasionally found favourable to the formation of seed ; evidently because the food that would have been conveyed into the branches, having no outlet, is forced into the fruit.

Another cause of sterility is the deficiency of pollen (87.) in the anthers of a given plant, as in vegetable mules (88.), which usually partake of the spermatic debility so well known in similar cases in the animal kingdom. It has often been found . that sterility of this kind is cured by the application, to the seedless plant, of the vigorous pollen of another less debilitated variety.

In some plants, such as Pelargoniums, when cultivated, the anthers shed their pollen before the stigma is ready to receive its influence, and thus sterility results. All such cases are provided for, by employing the pollen of another flower. (See Sweet in the *Gardener's Magazine*, vii. 206.)

An unfavourable state of the atmosphere obstructs the action of pollen, and thus produces sterility. Pollen will not produce its impregnating tubes in too low a temperature, or when the air is charged with moisture ; neither, in the absence of wind or insects, have some plants the power of conveying the pollen to the stigma, their anthers having no special irritability, and only opening for the discharge of the pollen, not ejecting it with force. If we watch the Hazel, or any of the

N

Coniferous order, in which the enormous quantity
of pollen employed to secure the impregnation of
the seed renders it easy to see what happens, it
will be found that no pollen is scattered in damp
cold weather; but, in a sunny, warm, dry morning,
the atmosphere surrounding such plants is, in the
impregnating season, filled with grains of pollen
discharged by the anthers. In wet springs the
crops of fruit fail, because the anthers are not
sufficiently dried to shrivel and discharge their
contents, which remain locked up in the anther
cells till the power of impregnation is lost. In
vineries and forcing-houses generally, into which
no air is admitted to disturb the foliage, nor any
artificial means employed for the same end,
and when the season is too early for the presence
of bees, flies, and other insects, the grapes will
not set: and in the frames of melons and cucumbers, from which insects are excluded, no seed
is formed unless the pollen is conveyed by hand,
from those flowers in which it is formed, to others
in which the young fruit alone is generated. In all
cases of this kind, the remedy for sterility is
obvious enough where plants exist in an artificial
condition; but, when they occur in the orchard or
the flower-garden, in the open air, science suggests
no assistance.

It sometimes happens that particular parts of
plants, distant from the fruit, are so constructed

as to attract to themselves the food intended for
the fruit, and thus to prevent the formation of seed.
For example : — The early varieties of Potato do
not readily produce seed, owing to the abstraction
by their tubers of the nutritive matter required for
the support of the seed. Mr. Knight found that by
destroying the tubers in part, as they formed, seeds
were readily procured from such varieties.

But perhaps the most frequent cause of sterility
is the monstrous condition of the flowers of many
cultivated plants. It has been fully explained
(84.) that the floral organs of plants are nothing
more than leaves, so modified as to be capable of
performing special acts, for particular purposes ; but
they are not capable of performing those acts any
longer than they retain their modified condition :
and therefore the stamens cannot secrete pollen,
when, by accidental circumstances, they are
changed into leaves, as happens in double flowers;
then, there is nothing to fertilise the stigma, and,
of course, no seed is produced. Or the carpels
themselves may be converted into leaves, and have
lost their seed-bearing property. Double flowers
in the latter case cannot possibly bear seed ; but
in the condition first mentioned they may, and
often do. To bring this about, the cultivator
plants in the vicinity of his sterile flowers others
of the same species, in which a part at least of the
stamens are perfect, and they furnish a sufficiency

of pollen for the impregnation of the other flowers in which there are no stamens.

In some cases, principally in those of Composite flowers, the seed is formed and advanced towards perfection, and then decays; this is owing to the flower heads of such plants being composed, in a great measure, of soft scales, absorbent and retentive of moisture, to which, in their own country, they are not exposed in the fruiting season, but by which they are affected under the hands of the cultivator. When the heads of such flowers are soaked with moisture, which they cannot get rid of, the scales rot, and decay spreads to all the other parts, and thus the production of seed is prevented. The Chinese Chrysanthemum is a familiar instance of this. Such plants seed readily if the flower heads are kept warm and dry; and it is thus that the sterile Chrysanthemum has been made seedful; that is to say, by growing it in a dry warm winter border, protected from showers by a roof of glass, or by using some such means of guarding it; or by rearing it in a warm dry climate.

When seeds are freely produced, it is not altogether a subject of indifference in what way they are saved, if it is desired that their progeny should be the most perfect that can be obtained. Weak seeds produce weak plants, and therefore recourse should be had, in all delicate cases, to artificial means for giving vigour to the seed. In general, the cultivator trusts to his eye for separating the plumpest and most completely formed seeds ; or to floating them in water, selecting only the heavy grains that sink, and rejecting all those which are buoyant enough to float. But the energy of the vital principle in a seed may be, undoubtedly, increased by abstracting neighbouring fruits, by improving the general health of the parent plant, by a full exposure of it to light, and by prolonging the period of maturation as much as is consistent with the health of the fruit. It is a constant rule that seedlings take after their parents, an unhealthy mother producing a diseased offspring, and a vigorous parent yielding a healthy progeny in all their minute gradations and modifications; and this is so true, that, as florists very well know, semi-double Ranunculi, Anemones, and similar flowers, will rarely yield double varieties, while the seed of the latter as unfrequently give birth to semi-double degenerations. Independently of these things, it is indispensable that the seed of a plant, when saved, should be perfectly ripe, if it is intended to be laid by for future sowing. The effect of ripening is to load the seed with carbon in the form of starch, or some such substance (102.), and to deprive it of water, conditions necessary for its preservation : but, if a seed is gathered before being ripe, these conditions are not secured; and, in pro-

portion to the deficiency of carbon and superabundance of water, is the seed liable to perish.

The complete maturation of the seed is, however,
a disadvantage, when it has to be sown immediately
after being gathered; for the embryo is formed, and
capable of germinating, long before the period of
greatest maturity. There are two periods in the
latter part of the organisation of a seed which,
although separated by no limits, require to be distinguished. The first is that when the embryo is
completed; and the second is when nature has, in
addition, furnished it with the means of maintaining its vitality for a long period. It is just as
capable of growing at the expiration of the first
period as of the second; it will do so immediately
if committed to the ground, and we see it actually happening to Peas, Beans, Corn, and other
field crops, in wet summers; but, at the end of
the second period, it cannot germinate till it has
relieved itself of all the carbon which, during that
period, was deposited in its tissue.

If seeds are to be preserved for a length of time,
a state of complete dryness is so necessary to them
that it has been recommended to increase it by
artificial means; not, however, by the application
of heat, or by any process like that of kiln-drying,
for that would destroy their vitality; but by some
of those chemical processes which dry the atmosphere without raising its temperature. It occurred

to Mr. Livingstone, that air made dry by means of sulphuric acid might be advantageously employed for this purpose, and he says that the success of his experiments was complete. He placed the seeds to be dried in the pans of Leslie's ice machine, and carefully replaced the receiver without exhausting the air; small seeds were sufficiently dried in one or two days, and the largest seeds in less than a week. (*Hort. Trans.*, iii. 184.)

Other contrivances might easily be adopted. Muriate of lime, for instance, which has the property of absorbing the moisture of the atmosphere, might, perhaps, be employed with advantage in drying the air in which seeds are placed after being gathered.

The reason why it is so important that seeds which have to be long kept should be thoroughly dried is, partly because seeds have the power of decomposing water, which causes the commencement of germination (14.), and, if this happens while they are cut off from the other means of existence, the process of growth must be stopped, and their death will follow; and, in part, from the tendency of vegetable matter in contact with water to putrefy, if the actions of life are not in play.

CHAP. VII.

OF SEED-PACKING.

IT seldom happens that seeds are sown as soon as they are ripe; it is sometimes desirable that they should be preserved for long periods of time; the power of conveying them for great distances, through various climates, is one of those upon which man most depends for the improvement of the horticultural resources of all countries; and for this purpose large sums are annually expended, both by governments and individuals. It is, therefore, an object of the first importance to ascertain what is not well understood, as it would seem, namely, the causes by which the destruction of the germinating power of seeds is effected; for it is only by doing this, that their preservation can be secured.

Seeds are probably possessed of different powers of life, some preserving their vital principle through centuries of time, while others have but an ephemeral existence under any circumstances. The reasons for this difference are unknown to us, and apparently depend upon a first cause, over which we have, therefore, no control; but the fact of great longevity in some seeds is certain, and there seems no reason why the conditions which enable them to preserve their germinating power for long periods of time should not be discovered and imitated.

Without admitting such doubtful cases as those of seeds preserved in mummies having germinated, there are many instances of seminal longevity about which there can be no doubt. Books contain an abundance of instances of plants having suddenly sprung up from the soil obtained from deep excavations, where the seeds must be supposed to have been buried for ages. Professor Henslow says that in the fens of Cambridgeshire, after the surface has been drained and the soil ploughed, large crops of white and black mustard invariably appear. Miller mentions a case of Plantago Psyllium having sprung from the soil of an ancient ditch which was emptied at Chelsea, although the plant had never been seen there in the memory of man. DeCandolle says that M. Gerardin succeeded in raising Kidneybeans from seed at least a hundred years old, taken out of the herbarium of Tournefort; and I have myself raised Raspberry plants from seeds found in an ancient coffin, in a barrow in Dorsetshire, which seeds, from the coins and other relics met with near them, may be estimated to have been sixteen or seventeen hundred years old.

In these cases, the only circumstances that we can conceive to have operated must have been such

a degree of dryness as prevented the decomposition
of the seed on the one hand, and the excitement of
its germinating powers on the other, a moderately
low temperature, and in some of them the exclu-
sion of air ; for moisture, heat, and communication
with the air, are necessary to enable seeds to grow
(14.). The tendency of moisture exposed to the
air, and in contact with inert vegetable matter,
such as a torpid seed, is by degrees to produce decay,
which rapidly spreads to the neighbouring parts.
But, if the vitality of a seed is excited by a fitting
temperature, the moisture with which it is in contact is then decomposed, the oxygen so obtained
combines with the carbon of the seed, and forms
carbonic acid which flies off, and by degrees re-
duces the amount of carbon lodged in the tissue of
the seed to that which is best suited for the growth
of the embryo (103.); then, if the embryo is so
situated that it cannot obtain from the surrounding
medium food upon which to subsist, its germination stops, and, being deprived of its carbon, the
safeguard of its vitality is removed, and it perishes.
If, however, the amount of moisture in contact
with a seed is very small, as in the dry earth at the
bottom of a tumulus for instance, the temperature
at the same time low, and the access of atmospheric
air cut off, neither putrefaction nor germination
is likely to occur. If seeds are exposed to a high
temperature in dryness, they will not perish, unless

the temperature rises beyond any thing likely to occur under natural circumstances. Edwards and Colin found that even wheat, barley, and rye, inhabitants of temperate countries, would bear when dry 104° for a long time without injury, although they died in three days in water at 95°; and a much higher prolonged temperature may be expected to produce no ill effect upon seeds inhabiting hotter countries. There is no apparent reason why the exposure of dry seeds to the air should destroy vitality, unless the exposure is very much prolonged; nor have we any evidence to show that it does, so long as they remain dry. The way in which the atmosphere would act injuriously upon dormant seeds is, by its oxygen abstracting their carbon; and it was formerly supposed that the carbonic acid extricated by germinating seeds was formed in this way. But the very valuable observations and experiments of Messrs. Edwards and Colin (See *Comptes rendus de l'Académie des Sciences*, vii. 922.) show that carbonic acid is formed by the assistance of the oxygen obtained by the decomposition of water.

If we apply these considerations to the plans usually employed for preserving artificially the vitality of seeds, we shall find them offer a ready explanation of the success that attends some methods of packing, and the constant failure of others.

The great object of those who have devised means

of packing seeds for distant journeys has, in general, been to exclude the air, and all other considerations have been subordinate to this. Enclosure in bottles hermetically sealed, in papers thickly coated with wax, in tin boxes, and similar contrivances, have been resorted to with this object in view : but no advantage can be derived from excluding the air, and the disadvantage is very great ; for the effect of excluding the air is to include whatever free moisture seeds may contain or be surrounded by ; this moisture is sufficient, in high temperatures, either to deprive the seed of its carbon of preservation, or to induce decay of the tissue, especially of the seed coats, which have no vitality themselves, and in either case the embryo perishes.

Packing in charcoal has been recommended, it is difficult to say why ; and experience shows what might have been anticipated, that it produces no other effect than packing in earth or other dry non-conducting material.

Clayed sugar has been employed, and, as it is said, occasionally with advantage ; but I have seen no instance of success, and, on the contrary, its tendency to absorb moisture from the air till it becomes capable of fermenting, is in itself an objection to the employment of this substance.

The most common method of packing is to enclose seeds in paper, to surround parcels of such papers with envelopes of the same material, and to enclose the whole in a deal box. It is in this manner that seedsmen usually despatch their orders to India, and other distant parts of the world. The evils of this method have been pointed out by Dr. Falconer, in the *Proceedings of the Horticultural Society*, vol. i. p. 49. "On one occasion," he says, " I received from England a large assortment of garden vegetable seeds, from a London seedsman. They were packed in the thick dark brown paper which is generally used by grocers and seedsmen, and which, for the facility of folding, is usually in a somewhat damp state. The packages were nailed up in a large wooden box, with numerous folds of this paper, and the box was then hermetically sealed in a tin case; it then found its way into the ship's hold. The damp paper, which in the temperature of England, say at 50°, would have mattered little, became an important agent when the ship got into the tropics ; at about 80° the damp became a hot vapour, and, when the seeds reached me, I found them all in a semi-pulpy and mildewed state."

Upon the whole, the only mode which is calculated to meet all the circumstances to which seeds are exposed during a voyage is, to dry them as thoroughly as possible, enclose them in coarse paper, and to pack the papers themselves very loosely in coarse canvass bags, not enclosed in boxes, but freely exposed to the air ; and to insure their transmission in some dry well-ventilated place. Thus, if the seeds are originally dried incompletely, they will become further dried on their passage; if the seed paper is damp, as it almost always is, the moisture will fly off through the sides of the bags, and will not stagnate around the seeds. It is true that, under such circumstances, the seeds will be exposed to the fluctuations of temperature, and to the influence of the atmosphere; but neither the one nor the other of these is likely to be productive of injury to the germinating principle. The excellence of this method I can attest from my own observation. Large quantities of seeds have been annually transmitted from India for many years, doubtless gathered with care, it is to be presumed prepared with every attention to the preservation of the vital principle, and certainly packed with all those precautions which have been erroneously supposed to be advantageous; the hopelessness of raising plants from such seeds has at length become so apparent, that many persons have altogether abandoned the attempt, and will not take the trouble to sow them when they arrive. But the seeds sent from India by Dr. Falconer, packed in the manner last described, exposed to all the accidents which those first mentioned can have encountered, have germinated so well, that we can scarcely say that the failure has been greater than if they had been collected in the south of Europe.

I have no doubt that the general badness of the

seeds from Brazil, from the Indian Archipelago, and from other intertropical countries, is almost always to be ascribed to the seeds having been originally insufficiently dried, and then enclosed in tightly packed boxes, whence the superfluous moisture had no means of escape.

For seeds containing oily matter, which are peculiarly liable to destruction (by their oil becoming rancid?), ramming in dry earth has been found advantageous; as in the case of the Mango.

CHAP. VIII.

OF PROPAGATION BY EYES AND KNAURS.

THE power of propagating plants by any other means than that of seeds depends entirely upon the presence of leaf-buds (*fig.* 16.), or, as they are technically called " eyes " (52.), which are in reality rudimentary branches in close connexion with the stem. All stems are furnished with such buds, which, although held together by a common system, have a power of independent existence under fitting circumstances; and, when called into growth, uniformly produce new parts, of exactly the same nature, with respect to each other, as that from which they originally sprang.

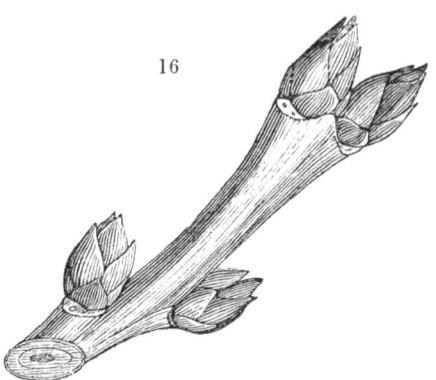

16

Under ordinary circumstances, an eye remains fixed upon the stem that generates it. There it grows, sending woody matter downwards over the alburnum, and a new branch upwards, clothed with leaves, and perhaps flowers : but it occasionally happens that eyes separate spontaneously from their mother stem, and when they fall upon the ground they emit roots and become new plants (p. 32. fig. 3.). This happens in several kinds of Lily, and in other genera.

Man has taken advantage of this property, and has discovered that the eyes of many plants, if separated artificially from the stem and placed in earth, will, under favourable circumstances, produce new plants, just as such eyes would have done if they had spontaneously disarticulated ; hence the system of propagation by eyes, an operation employed only to a limited extent in actual practice, but which in theory seems applicable to all plants whatever. The only species very generally so increased

are the Potato and the Vine. Of the latter, the
eye, with a small portion of the stem adhering to
it, is commonly taken as the means of obtaining
young plants ; being placed in earth, with a bottom
heat of 75° or 80°, and kept in a damp atmosphere,
it speedily shoots upwards into a branch, and at
the same time establishes itself in the soil by the
developement of the requisite quantity of roots.
In order to insure success in this operation upon
the Vine, it is only necessary that the eye should
be dormant, and that a small piece of well-ripened
wood should, as has been already stated, be separated with it ; it will then grow in much the same
way, and under the same circumstances as a seed.
There is no doubt that many plants could be
thus multiplied just as easily as the Vine, but it
is equally certain that a far larger number cannot
be so increased. The reason of this is, probably,
that such eyes are not sufficiently excitable, and
that consequently they decay before their vital
energies are roused ; and, in addition, that they do
not contain within themselves a sufficient quantity
of organisable matter upon which to exist until
new roots are formed, and capable of feeding the
nascent branch.

Mr. Knight's explanation of this, although in
part applicable to cuttings only, yet seems to deserve being introduced in this place. " Every
leaf-bud is well known to be capable of extending

o

itself into a branch, and of becoming the stem of a future tree ; but it does not contain, nor is it at all able to prepare and assimilate, the organisable matter required for its extension and developement. This must be derived from a different source, the alburnous substance of the tree, which appears the reservoir, in all this tribe of plants, in which such matter is deposited.   I found a very few grains of alburnum to be sufficient to support a bud of the Vine, and to occasion the formation of minute leaves and roots : but the early growth of such plants was extremely slender and feeble, as if they had sprung from small seeds ; and the buds of the same plant, wholly detached from the alburnum, were incapable of retaining life.   The quantity of alburnum being increased, the growth of the buds increased in the same proportion ; and when cuttings of a foot long, and composed chiefly of two-years-old wood, were employed, the first growth of the buds was nearly as strong as it would have been, if the cuttings had not been detached from the tree.   The quantity of alburnum in every young and thriving tree, exclusive of the Palm tribe, is proportionate to the number of its buds; and if the number of these were, in any instance, ascertained and compared with the quantity of alburnous matter in the branches and stem and roots, it would be found that nature has always formed a reservoir sufficiently extensive to supply every bud.   But those of a cutting, under the most favourable circumstances, must derive their nutriment from a more limited and precarious source ; and it is therefore expedient that the gardener should, in the first instance, make the most ample provision conveniently within his power for their maintenance, and that he should subsequently attend very closely to the economical expenditure of such provision." (*Horticultural Transactions*, ii. 115.)

In the Potato the requisite provision of organisable matter is always secured, in consequence of the great difficulty of separating an eye of that plant, without fragments of the amylaceous tuber adhering to it.

The provision of alimentary matter may, however, be, in some cases, disadvantageous, by promoting too great a developement of stems and leaves, of which the Potato itself is an instance. Theoretically, the more nutritive matter there is for the eyes, the greater crop there will be, *cæteris paribus*, and so there probably is of leaves and stems; and it would seem that whole potatoes should be more advantageous to plant than sets. But I have proved by a series of numerous experiments, that the weight of potatoes per acre is greater, under equal circumstances, from sets than from whole tubers, by upwards of from seven cwt. to three tons per acre; and considerably more, on comparison of the clear produce, after deducting the weight of sets employed in both cases. (*Hort. Trans.*, n. s., i. 445. and ii. 156.) In these instances, I supposed the rankness of the vegetation from the whole tubers to be the cause of the diminished crop, for the stems were unable to support themselves, and were blown about, laid, and broken by the wind.

While, however, in such plants as the Potato, all the eyes are equally capable of forming new tubers, it is found by experience that they do so with different degrees of rapidity, according to the age of that part of the stem or tuber which furnishes them. It is stated by a writer in the *Gardener's Magazine* (vol. i. p. 406.), that it is well known in Lancashire to some cultivators of the Potato, "that different eyes germinate and give their produce, or become ripe, at times varying very materially, say several weeks, from each other; some being ripe or fit for use as early as the middle of May, and others not till June or July. This will be understood by reference to the accompanying sketch. The sets nearest the extremity of the Potato (*fig.* 17. *a*) are *soonest ripe*, and, in Lancashire, are planted in warm places in March or the beginning of April, and are ready for the market about the 12th or 15th of May. The produce of the next sets (*b*) is ready in about a fortnight after, and that from the root end (*c* and *d*) still later. These root end sets (from *b* to *d*) are usually put together, and the extremity of the root end is thrown aside for the pigs." This fact, if correctly stated, shows, not that the youngest eyes, or those nearest the point of the Potato, are the ripest, which is impossible, but that they are more excitable, and consequently grow more rapidly than those at the middle or base.

Besides the cases of propagation by eyes now mentioned, there is another of which a notice is given by Signor Manetti (*Gardener's Magazine*, vii. 663.), as practised in Italy upon the Olive. It appears that, from old Olive trees, certain knots or excrescences, called *uovoli*, are cut out of the bark, of which a portion is left adhering to them, and, being planted, grow into young Olive trees. Of these I have no further account ; but it is evident that the *uovoli* are no other than our knaurs (*fig.* 18.), already spoken of (53.) under the name of *embryo buds;* concretions found in the bark of many, and probably of all, trees, and supposed to have been adventitious buds developed in the bark, and, by the pressure of the surrounding parts, forced into those tortuous woody masses in the shape of which we find them. They, in general, present an oblong or conical form, are sometimes collected into clusters (*fig.* 18. *a*), and usually exhibit little or no appearance of a tendency to further growth.   It is, however, not uncommon to find them lengthening into branches, as is shown in the

o 3

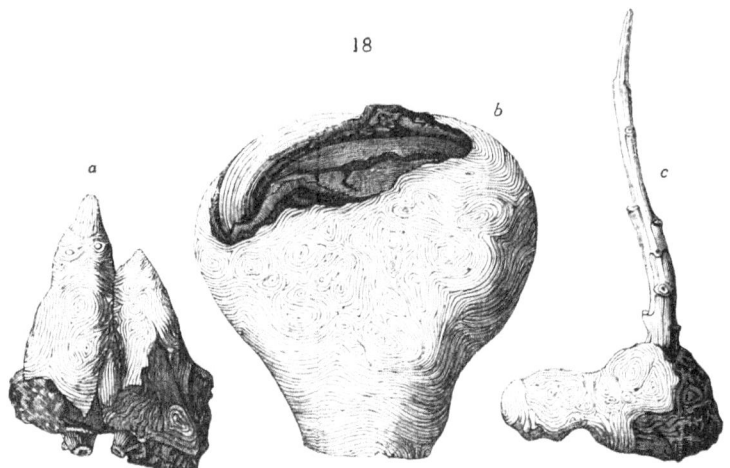

Poplar (*fig*. 18. *c*), for which I am indebted to
Sir Oswald Mosley ; and although they have never
yet been used for the purposes of propagation,
except in the case of the Olive, there seems to be
no reason why they should not be so employed, if
any necessity were to arise for them. The real
amount of their powers of growth is unknown, and
would be a good subject of investigation.

CHAP. IX.

OF PROPAGATION BY LEAVES.

In the beginning of the last century, Richard
Bradley, a Fellow of the Royal Society, published
a translation from the Dutch of Agricola, of a book

upon the propagation of plants by leaves, in which it was asserted that, by the aid of a mastic invented by the author, the leaves of any plant, dipped at the stalk end into this preparation, would immediately strike root ; and the book was adorned with copperplates exhibiting both the process and its result, in the form of fields stuck full of Orange leaves growing into trees.

Although this work was very absurd, yet it probably originated in the discovery that the leaves of some plants will grow under special circumstances ; a fact often supposed to be much more rare than it really is. In Professor Morren's French translation of my *Outlines of the First Principles of Horticulture*, Rochea falcata* is named as producing adventitious buds (53.) from the upper side of its leaves ; and the Orange, the Aucuba, and the Fig, as other instances of leaves which will multiply their species (p. 152.): the power of Bryophyllum to do the same thing is familiar to every one. Hedwig found the leaves of the Crown Imperial, put into a plant-press, produce bulbs from their surface. There is a well-known case of the same effect having been observed in Ornithogalum thyrsoideum. Mr. Auguste de St. Hilaire mentions an instance of leaf-buds generated by fragments of the leaves of " Theophrasta," which had been buried

* See, also, DeCandolle's Physiologie Végétale, ii. 672.

by M. Neumann, chief gardener at the Garden of Plants at Paris, and of young Droseras furnished by the leaves of Drosera intermedia. Mr. Henry Cassini is said to have seen young plants produced by the leaves of Cardamine pratensis; English botanists know that offsets spring from the margins of the leaves of Malaxis paludosa; in our stoves we see Ferns of many kinds, especially Woodwardia radicans, propagating themselves by offsets from the leaves; Mr. Turpin tells us that floating fragments of Watercress leaves, cut up by a species of Phryganea for its nest, "produce presently from their base, and below the common petiole, at first two or three colourless roots, then in their centre a small conical bud, green, in which are found, or rather from which successively arise, all the aërial parts of a new Watercress plant, while the roots multiply and lengthen." (*Comptes rendus*, 1839, sem. 2., 438.) Mr. Flourens also mentions a case of Purslane, whose leaves, divided into three, produced as many new plants, each having a root, stem, and leaves. In the *Transactions of the Horticultural Society*, is an account of a Zamia each of whose scales (*fig.* 19.) produced a new plant, when the central part of the stem was decayed. Finally, the following case is named in the same work (vol. v. p. 242.) by Mr. Knight :—

"In an early part of the summer, some leaves of Mint (Mentha piperita), without any portion of

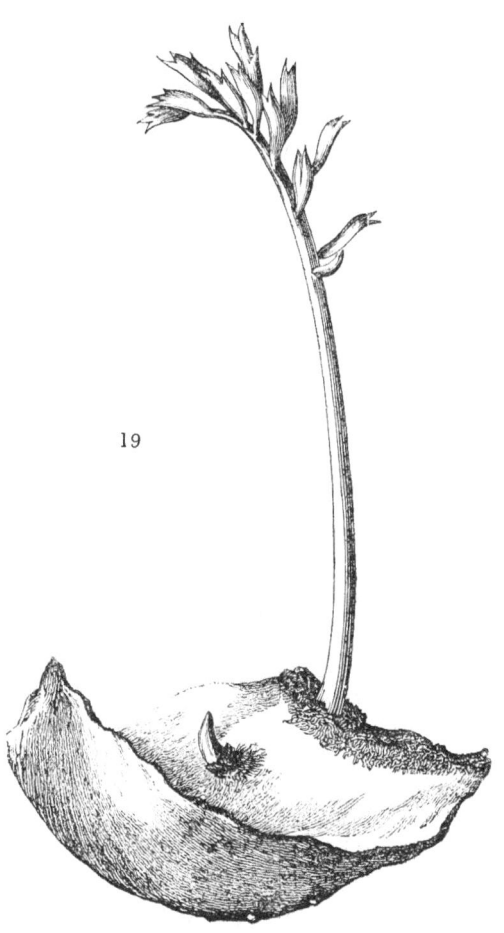

19

the substance of the stems upon which they had grown, were planted in small pots, and subjected to artificial heat, under glass. They emitted roots, and lived more than twelve months, having assumed nearly the character of the leaves of evergreen trees; and upon the mould being turned out of the pots, it was found to be everywhere surrounded by just such an interwoven mass of roots, as would have been emitted by perfect plants of the same species. These roots presented the usual character of those organs, and consisted of medulla, alburnum, bark, and epidermis; and as the leaf itself, during the growth of these, increased greatly in weight, the evidence that it generated the true sap which was expended in their formation appears perfectly conclusive."

In our gardens, we know of many other cases of the same kind.   Hoya is a common instance, and three others are here figured (*fig.* 20.); viz., Gesnera (*a*), Clianthus puniceus (*b*), Gloxinia speciosa (*c*).   In these, and all such cases, the first thing that happens is an excessive developement of cellular tissue, which forms a large convex "callus" at the base; from which, after a time, roots proceed; and by which eventually a leaf-bud, the commencement of a new stem, is generated.

It is not surprising that leaves should possess this quality, when we remember that every leaf does the same thing naturally, while attached to the plant that bears it; that is to say, forms at its base a bud which is constantly axillary to itself. Leaves, however, have not been often employed as the means of propagating a species; and it is probable that most leaves, when separated from their parent,

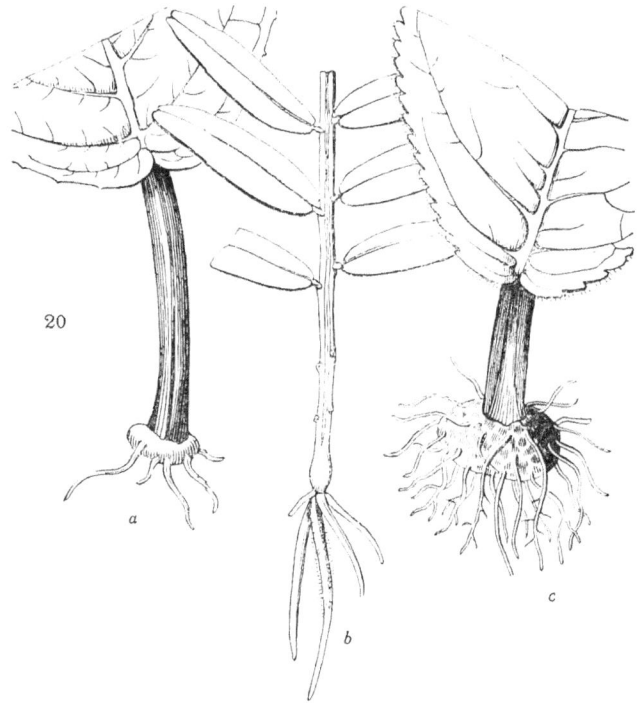

are incapable of doing so, for reasons which we are
not as yet able to explain. The most common case
of their employment is in the form of the scales
of a bulb, which will, with some certainty, produce
new plants under favourable circumstances. Those
circumstances are, a strong bottom heat, moderate
moisture, and a rich stimulating soil.

When plants are produced by leaves under ordinary circumstances, the conditions most favourable
to their doing so are of the same nature. A moderate amount of moisture prevents their dying

from perspiration or perishing from decay; a good bottom heat stimulates their vital forces, and causes them to exercise whatever power they possess; and, in addition, they are covered by a slightly shaded bell glass, which maintains around them an atmosphere of uniform humidity, and, at the same time, cuts off the approach of those direct solar rays, which, acting as a stimulus to perspiration, would have a tendency to exhaust the leaves of their fluid before they could organise, at their base, the new matter from which the leaf-bud is eventually produced.

CHAP. X.

OF PROPAGATION BY CUTTINGS.

THIS, which is the most common of all modes of artificial propagation except grafting, depends upon essentially the same principle as propagation by eyes; that is to say, the pieces of a plant called cuttings possess a power of growth in consequence of their bearing leaf-buds or eyes upon their surface. In striking by eyes, we have the great difficulty to encounter of keeping the eye active till it has organised roots with which to feed itself; the

earth furnishes such a supply unwillingly or unsuit-
ably, nature intending that the bud should, in the
first instance, be supported by the soluble nutriment ready prepared and lodged in its immediate
vicinity, in the pith or some other part of the
stem.   For this reason, cuttings, which consist of
eyes and the part containing their proper aliment,
usually strike root more freely than eyes by themselves.

This being so, it is plain that a cutting is only
capable of multiplying a plant when it bears buds
upon its surface ; and as the stem is the only part
upon which buds certainly exist, so the stem is the
only part from which cuttings should be prepared.
And again, as the internode, or that space of the
stem which intervenes between leaf and leaf, has
no buds, their station being confined to the axil of

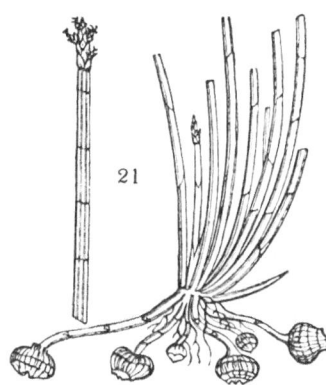

the leaves, a cutting prepared from an internode
only is as improper as one
from the root.   It is no
doubt true, that we con-
stantly propagate plants
from pieces of what are
called roots, as in the Potato, or the Scirpus tube-
rosus (*fig.* 21.); but such
roots are, in reality, the kind of stem called a
tuber (51.) : and, in like manner, other cases of

similar propagation are also successful, because the part called a root is, in reality, an underground stem covered with the rudiments of leaves, to each of which an eye belongs. The Rose, the Lilac, and many other plants have subterranean stems, cuttings of which will therefore answer the purpose of propagation. It will also occasionally happen, that, owing to unknown causes, morsels of the true root will generate what are called adventitious buds; and hence we do occasionally see the root employed for propagation, as in Cydonia japonica; but these are rare and exceptional cases, and by no means affect the general rule. Mr. Knight attempted to account for this, by supposing that the powers which roots of various forms, and cuttings, and other detached parts of plants, possess of emitting foliage, "are wholly, and in all cases, dependent upon the presence of true sap previously deposited within them." (*Hort. Trans.*, v. 242.) But this is a very obscure expression, and does not seem to throw any light upon the subject.

When the Vine grows in a very warm damp stove, its stem emits roots into the air; the same thing happens to the Maize on the lower part of its stem; and in these and all such cases, the roots are found to be emitted from buds. Hence it has been inferred that the roots of a plant are as much productions of buds as branches are, and that the stem is nothing more than a collection of such roots held together under the form of wood and bark. The present is not the place for a renewal of this discussion, for the arguments in favour of and opposed to which, the reader is referred to my *Introduction to Botany*, 3d edit. p. 309. &c. It is sufficient here, to remark that the question turns upon whether the buds and leaves actually themselves produce roots, or merely furnish the organisable matter out of which roots are formed ; and that, therefore, for the purpose of horticulture, either the one or the other is equally capable of explaining the facts connected with cuttings.*

* The following curious fact, recorded by Mr. Livingstone, which seems to have escaped observation, deserves to be mentioned here : — " The Pterocarpus Marsupium, one of the most beautiful of the large trees of the East Indies, and which grows in the greatest perfection about Malacca, affording, by its elegant wide-expanding boughs, and thick-spreading pinnated leaves, a shade equally delightful with the far-famed Tamarind tree, is readily propagated by cuttings of all sizes, if planted even after the pieces have been cut for many months, notwithstanding they appear quite dry, and fit only for the fire. I have witnessed some of three, four, five, six, and seven inches in diameter, and ten or twelve feet long, come to be fine trees in a few years. While watching the transformation of the log into the tree, I have been able to trace the progress of the radicles from the buds, which began to shoot from the upper part of the stump a few days after it had been placed in the ground, and marked their progress till they reached the earth. By elevating the bark, minute fibres are seen to descend contemporaneously as the bud shoots into a branch. In a few weeks these are seen to interlace each other. In less than two years the living

As far as physiology can explain the operation of propagation by cuttings, it appears that roots are formed by the action of leaves; that branches are developed from the buds; and that the buds are maintained by the suitable aliment stored up in the stem. Every thing beyond this seems to be connected with specific constitutional powers, of which science can give no explanation.

In considering what conditions are most favourable to the maintenance of a cutting in the state required, in order to enable it to become a young plant, it will be most convenient, in the first place, to examine the rationale of some one method which is known to be successful. For this purpose, the following detail, by Mr. Knight, of his mode of striking the Mulberry, is selected : —

"A considerable number of cuttings were taken from the most vigorous bearing branches of a Mulberry tree, in the middle of November, 1812, and were immediately reduced to the length adapted to small pots, in which I proposed them ultimately to be planted, and which were between four and five inches deep. Each cutting was composed of about two parts of two-years-old wood, that is, wood of

---

fibrous system is complete ; in five years no vestige of its log origin can be perceived; its diameter and height are doubled, and the tree is in all respects as elegant and beautiful as if it had been produced from seed." (*Hort. Trans.*, iv. 226.)

the preceding year, and about one third of yearling wood, the produce of the preceding summer ; and the bottom of each was cut so much aslope, that its surface might be nearly parallel with that of the bottom of the pot in which it was to be placed.

" The cuttings were then inserted in the common ground, under a south wall, and so deeply immersed in it, that one bud only remained visible above its surface ; and in this situation they remained till April. At this period the buds were much swollen, and the upper ends of the cuttings appeared similar to those of branches which had been shortened in the preceding autumn, and become incapable of transmitting any portion of the ascending fluid. The bark at the lower ends had also begun to emit those processes which usually precede the production of roots. The cuttings were now removed to the pots to which they had been previously fitted, and placed in a moderate hot-bed ; a single bud only of each cutting remained visible above the mould, and that being partially covered ; and in this situation they vegetated with so much vigour, and emitted roots so abundantly, that I do not think one cutting in a hundred would fail, with proper attention.   Some of the pots were placed round the edges of a melon bed, which affords a very eligible situation where a few plants only are wanted." (*Horticultural Transactions*, ii. 117.)   In this case success appears to have depended upon the following circumstances :—

1. The cuttings were prepared in November, at the end of the season of growth, when all the organisable matter required for the cutting was formed, and locked up in the proper places in its interior. It was not necessary, therefore, to take any means of insuring a further supply of aliment. But had it been otherwise, that is to say, if the cuttings had been prepared in the summer, in the midst of their growth, it would have been indispensable to allow a leaf or two to remain attached to the upper end of the cutting, to assist in the formation of alimentary matter.

2. Although but one eye was allowed to grow, yet the cuttings themselves were four or five inches long, and they consisted, to the extent of two thirds, of two-years-old wood. By this means the quantity of food for the nascent branch was intended to be so great as to insure it against suffering from an inadequate supply, until it had formed roots. The importance of this has already been shown by Mr. Knight in a previous part of this Book.

3. The cuttings were taken off in November, and not in the spring. This gave them time to form granulations of cellular substance at the lower or wounded end, before the powers of absorption by the alburnum were aroused, and so to protect themselves against a too copious supply of aqueous matter before the growing bud could dispose of it

by its leaves (64.). This protection is afforded by the thinnest stratum of new cellular tissue, which covers over the ends of the wounded vessels, and acts as a vital filter through which all the crude food must pass from the soil.

4. The lower end of the cuttings was so divided as to be parallel with the bottom of the pot, and it appears from the context, although it is not expressly so stated, that this end was to *touch* the bottom of the pot. The importance of this precaution is well known ; cuttings of the Lemon and Orange, which are by no means willing to strike if it is neglected, become young plants readily if it is attended to ; and in all difficult cases it is had recourse to. The object of it seems to be to place the absorbent or root end of the cutting in a situation where, while it is completely drained of water, it may, nevertheless, be in the vicinity of a never-failing supply of aqueous vapour. If it were surrounded by earth, water would readily collect about it in a condensed state, and, the vessels being all open in consequence of being cut through, would rise at once into the interior; but the application of the root end immediately to the earthen bottom of the pot, with which it is so cut as to be nearly parallel, necessarily prevents any such accumulation and introduction of water, unless over-watering is allowed, and this all good gardeners will take care to avoid. An ingenious plan of Mr. Forsyth's is intended

to answer this purpose rather more perfectly. He

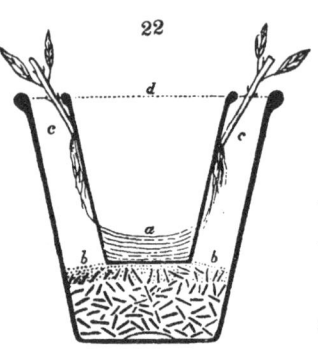

puts a small sixty pot (*fig.* 22. *a d*) into one of larger size, having first closed up the bottom of the former with clay (*a*); then having filled the bottom of the outer pot with crocks, he fills up the sides (*c c*) with propagating soil, in which his cuttings are so placed that their root ends rest against the sides of the inner pot; the latter is then filled with water, which passes very slowly through the sides until it reaches the cuttings. (See *Gardener's Magazine*, vol. xi. p. 564.) In many cases, especially in striking such plants as Heaths, gardeners employ a stratum of silver sand, placed immediately over the earth in which such plants love to grow. The cuttings are inserted into the sand, but so near the earth that the roots, presently after their emission, find themselves in it, and consequently in contact with a source of food. This sand answers the same purpose as placing the root end of the cutting in contact with the pot, and is an ingenious device for doing that with small cuttings, which cannot be conveniently done otherwise except with large ones.

5. The cuttings are eventually placed in a hot-

bed.   This is for the purpose of giving them a stimulus at exactly that time when they are most
ready to receive it.   Had they been forced at first
in bottom heat, the stimulus would have been applied to cuttings whose excitability had not been
renovated (113.), and the consequence would have
been a developement of the powers of growth so
languid, that they probably would not have sur-
vived the coming winter : but, the stimulus being
withheld till the cuttings are quite ready for growth,
it tells with the utmost possible effect.

In addition to these comments upon an excellent
mode of striking cuttings of many kinds, it is necessary to add some observations upon the object of
additional precautions often taken by gardeners.

Cuttings are covered by bell glasses, whose edge
is pressed into the earth.   This is for the purpose
of preserving a uniform degree of humidity in the
atmosphere breathed by the cuttings.   It is generally
necessary to leave one or more leaves upon a cutting, in order to generate organisable matter, and
to assist in the formation of roots ; but this is a
very delicate operation, for, if the leaf is allowed
to suffer by excessive perspiration, the cuttings
must necessarily perish (75.).   To maintain a steady
saturated atmosphere around a cutting stops this
danger, and hence the use of a bell glass. A double
glass has even been recommended (*fig.* 23.) ; but,
if this precaution is of any value, it must be, not

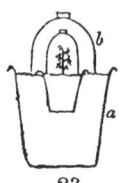

because it maintains an even temperature, which is injurious rather than
useful, but because it prevents condensation upon the inner bell glass, and the
consequent abstraction of atmospheric
moisture, and probably acts at the same time as a
kind of shade.

Notwithstanding the precaution of covering cuttings with a bell glass, shade is also necessary, as a
further security against perspiration ; for light acts
as a specific stimulus (71.), whose effects are very
difficult to counteract. It must, however, be employed with great caution ; for, if there is not light
enough, the leaves attached to the cuttings cannot
form that organisable matter out of which roots are
produced.

All gardeners know that the root end of a cutting should be *close below* a leaf-bud ; this is to
facilitate the emission of roots by the buds, which
emission must necessarily take place with greater
or less difficulty in proportion as their exit is facilitated or impeded by the pressure of bark on them.

No further precautions are taken with cuttings,
nor does it at first sight appear possible to suggest
any : nevertheless, the enormous constitutional difference among plants is such, that, while numerous
species will strike without any difficulty under almost any circumstances, with the wood ripe or

half-ripe, just formed or aged, there are many others which no art has yet succeeded in converting into plants; and it is by no means uncommon to find that, out of a potful of cuttings of the same species, apparently all alike and subjected to exactly the same treatment, one will grow and the remainder fail.   It is, therefore, one of the most probable of all things, that the principles of striking cuttings are still very imperfectly understood, and that this is one of the points of horticulture in which there is the greatest room for improvement.

It may be worthy of enquiry whether bell glasses of different colours will not produce different effects upon cuttings, in consequence of their different power of transmitting light.   It has been shown by Dr. Daubeny, in a very interesting paper in the *Philosophical Transactions* for 1836, page 149., that glass of different colours exercises very different effects upon the plants exposed to the rays of solar light passing through it ; that both the exhalation and absorption of moisture by plants, so far as they depend upon the influence of light, are affected in the greatest degree by the most luminous rays, and that all the functions of the vegetable œconomy, which are owing to the presence of this agent, follow in that respect the same law.   In these experiments it was ascertained that the glass employed admitted the passage of the rays of light in the following proportions : —

	Transparent.	Orange.	Red.	Blue.	Purple.	Green.
Luminous rays	7	6	4	4	3	5
Calorific rays	7	6	5	3	4	2
Chemical rays	7	4	0	6	6	3

M. Decaisne found, during some experiments to ascertain the effect of light in causing the production of colouring matter in the Madder plant, that when the lower parts of a plant were enclosed in cases glazed at the side with transparent green, red, or yellow glass, the leaves and stem of the part surrounded by red glass became pallid, and exhibited signs of suffering in a greater degree than under the other colours, but all were affected more or less.* (*Recherches sur la Garance*, p. 23.)

We however require very different experiments from any yet instituted, before we can proceed to draw practical conclusions as to the relative effects upon plants of glass of different colours. We do not know what the effect is of the calorific and chemical rays, and therefore we cannot say what may be the advantage or disadvantage of their action upon plants. As, however, the object of the cultivator is to protect his cuttings from too much light, and at the same time to give them enough to enable them to perform their digestive functions steadily, there can be little doubt that transparent glass is inferior to that of another colour.

* The nature of these experiments has been misapprehended in the translation, by Mr. Francis, of Meyen's *Report on Vegetable Physiology* for 1837, p. 51.

CHAP. XI.

OF PROPAGATION BY LAYERS AND SUCKERS.

WITH regard to layers, there is but little which it is necessary to say regarding them, if what has been stated respecting eyes, leaves, and cuttings, has been rightly understood and well considered. A layer is a branch bent into the earth, and half cut through at the bend, the free portion of the wound being called "a tongue." It is, in fact, a cutting only partially separated from its parent.

The object of the gardener is to induce the layer to emit roots into the earth at the tongue. With this view he twists the shoot half round, so as to injure the wood-vessels; he heads it back so that only a bud or two appears above ground ; and, when much nicety is requisite, he places a handful of silver sand round the tongued part ; then pressing the earth down with his foot, so as to secure the layer, he leaves it without further care. The intention of both tongueing and twisting is to prevent the return of sap from the layer into the main stem, while a small quantity is allowed to rise out of the latter into the former ; the effect of this being to compel the returning sap to organise itself externally as roots, instead of passing downwards below the bark as wood. The bending back is to assist in this object, by preventing the expenditure of sap in the formation or rather completion of leaves; and the silver sand is to secure the drainage so necessary to cuttings.

In most cases, this is sufficient; but it must be obvious that the exact manner in which the layering is effected is unimportant, and that it may be varied according to circumstances. Thus, Mr. James Munro describes a successful method of layering brittle-branched plants by simply slitting the shoot at the bend, and inserting a stone at that place (*Gardener's Magazine*, ix. 302.): and Mr. Knight found that, in cases of difficult rooting, the process is facilitated by ringing the shoot just below the tongue, about midsummer, when the leaves upon the layers had acquired their full growth (*Hort. Trans.*, i. 256.); by which means he prevented the passage of the returning sap further downwards than the point intended for the emission of roots.

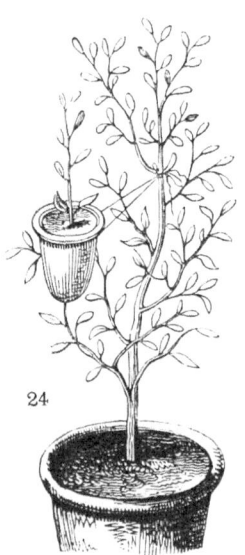

24

It will sometimes happen that the branch of a plant cannot be conveniently bent downwards into the earth; in such cases, the earth may be elevated to the branch by various contrivances, as is commonly done by the Chinese. (*fig.* 24.) When this is done, no other care is necessary than that required for layers, except to keep the earth surrounding the branch steadily moist.

Suckers are branches naturally thrown up by a plant from its base, when the onward current of growth of the stem is stopped. Every stem, even the oldest, must have been once covered with leaves; each leaf had a bud in its axil; but, of those buds, few are developed as branches, and the remainder remain latent or perish. When the onward growth of a plant is arrested, the sap is driven to find new outlets, and then latent buds are very likely to be developed; in fact, when the whole plant is young, they must necessarily shoot forth under fitting circumstances; the well-known effect of cutting down a tree is an exemplification of this. Such branches, if they proceed from under ground, frequently form roots at their base, when they are employed as a means of propagation; and, in the case of the Pine-apple, they are made use of for the same purpose, although they do not emit roots till they are separated from the parent. Gardeners usually satisfy themselves with taking from their Pine-apple plants such suckers as are produced in consequence of the stoppage of onward growth by the formation of the fruit : but these are few in number, and not at all what the plant is capable of yielding. Instead of throwing away the " stump" of the Pine-apple, it should be placed in a damp pit, and exposed to a bottom heat of 90 or thereabouts, when every one of the latent eyes will spring forth, and a crop of young plants be the result. Mr. Alexander Forsyth, a very sensible writer upon these subjects, pointed this out some years since in the *Gardener's Magazine* (xii. 594.) ; and there can be no doubt that his observations upon the folly of throwing away stumps are perfectly correct both in theory and practice.

The practice of scarring the centre of bulbs, the heads of Echinocacti and such plants, and the crown of the stem of species like Littæa geminiflora, in all which cases suckers are the result, is explicable upon the foregoing principles.

CHAP. XII.

OF PROPAGATION BY BUDDING AND GRAFTING.

THESE operations consist in causing an eye or a cutting of one plant to grow upon some other plant, so that the two, by forming an organic union, become a new and compound individual. The eye, in these cases, takes the name of bud, the cutting is called a scion, and the plant upon which they are made to grow is named the stock.

Propagation by eyes and cuttings is, therefore, the same as budding and grafting, with this important difference, that in the one case the fragments of a plant are made to strike root into the inorganic soil, and to grow on their own bottom, as the saying is, while in the other they emit the equivalent of roots into living organic matter. In like manner, the operation of inarching, or causing the branch of one plant to remain attached to its parent, and at the same time to grow upon the branch of another tree, is analogous to layering.

The objects of these operations are manifold. Many plants, such as the Pear and the Apple, will bud or graft freely, but are difficult to strike from cuttings. Species which are naturally delicate become robust when " worked " on robust stocks; and the consequence is a more abundant production of flowers and fruit : thus the more delicate kinds of Vines produce larger and finer grapes when worked upon such coarse robust sorts as the Syrian and Nice. The Double yellow Rose, which so seldom opens its flowers, and which will not grow at all in many situations, blossoms abundantly, and grows freely, when worked upon the common China Rose. (See *Hort. Trans.*, v. 370.) The peculiar qualities of some plants can only be preserved by working : this is especially the case with certain kinds of variegated Roses, which retain their gay markings when budded, but become plain if on their own bottom. (*Ib.*, 492.) Fruit may be obtained from seedling plants by these processes much earlier than by any others, and thus many years' uncertain expectation may be saved : indeed, Mr. Knight ascertained that it is possible to transfer the blossom-buds of one plant to another, so as to obtain flowers or fruit from them immediately. He thus fixed on the Wild Rose the flower-buds of Garden Roses, "and these buds, being abundantly supplied with nutriment, afforded much finer roses than they would have done had they retained their natural situation." He repeated many similar experiments upon the Pear and Peach trees with similar success ; but, in the case of the Pear, he found that if the buds were inserted earlier than the end of August or beginning of September, they became branches and not flowers.

The manner in which these operations may be practised is exceedingly various, and an abundance of fanciful methods have been devised, for an account of which the reader is referred to Thouin's *Monographie des Greffes ;* to the article "Greffe " by the same author, in the *Nouveau Cours complet d'Agriculture,* &c., edition of 1822 ; to Loudon's

Encyclopædia of Gardening, part ii. ; and to the *Gardener's Magazine*, vol. x. p. 305.   I shall only here describe the commoner and more important methods.

BUDDING consists in introducing a bud of one tree, with a portion of bark adhering to it, below the bark of another tree.   In order to effect this, a longitudinal incision is made through the bark of

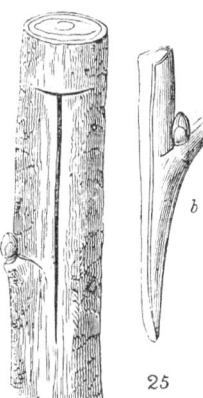

the stock down to the wood, and is then crossed at the upper end by a similar cut (*fig. 25. a*), so that the whole wound resembles the letter T.   Then from the scion is pared off a bud with a portion of the bark (*fig. 25. b*), and the latter is pushed below the bark of the stock until the bud is actually upon the naked wood of the stock; the upper lips of the wound in the stock and that of the bud are made to coincide, the whole are fastened down by a ligature, and the operation is complete.

By these means we gain the important end of bringing in close contact a considerable surface of young organising matter.   The organisation of wood takes place on its exterior, and that of bark on its interior surface, and these are the parts which are applied to each other in the operation

of budding ; in addition to which the stranger bud
finds itself, in its new position, as freely in communication with alimentary matter, or more so, than
on its parent branch.   A union takes place of the
cellular faces, or horizontal system, of the stock and
bark of the bud, while the latter, as soon as it begins to grow, sends down the woody matter, or
vertical system, through the cellular substance.   In
consequence of the horizontal incision, the returning sap of the scion is arrested in its course, and
accumulates a little just above the new bud, to
which it is gradually supplied as it is required.
Sometimes the whole of the wood of the bud below
the bark is allowed to remain; and, in that case,
contact between the organising surfaces of the
stock and scion does not take place, and the union
of the two is much less certain : as it is, however,
usually practised with tender shoots before the
wood is consolidated, the contact spoken of is of
less moment.   In all cases, a portion of the wood
of the bud must be left adhering to it, or the bud
will perish ; because its most essential part is the
young woody matter in its centre, and not the
external surface, which is a mere coating of bark.

In the *Agricultural Journal of the Pays Bas* for
October, 1824, it is recommended to reverse the
usual mode of raising the bark for inserting the
buds, and to make the cross cut at the bottom of
the slit, instead of at the top, as is generally done

in Britain. The bud is said rarely to fail of success, because it receives abundance of the descending sap, which it cannot receive when it is under the cross cut. This explanation is unintelligible, and there is no apparent advantage in the method; it is, however, practised by the orange-growers of the South of France.

Mr. Knight was accustomed on some occasions to employ two distinct ligatures to hold the bud of his Peach trees in its place. One was first placed above the bud inserted; and upon the transverse section through the back: the other, which had no further office than that of securing the bud, was employed in the usual way. As soon as the bud had attached itself, the ligature last applied was taken off: but the other was suffered to remain. The passage of the sap upwards was in consequence much obstructed, and buds inserted in June began to vegetate strongly in July: when these had afforded shoots about four inches long, the remaining ligature was taken off to permit the excess of sap to pass on; and the young shoots were nailed to the wall. Being there properly exposed to light, their wood ripened well, and afforded blossoms in the succeeding spring.

Flute-budding (*fig.* 26.) is not practised in this country, but deserves mention. It consists of peeling off a ring of bark from the stock, just below a terminal bud; replacing it by a similar ring, with a

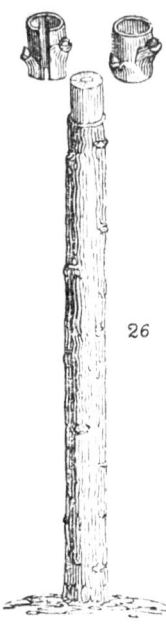

bud or two upon it, taken from a scion; and binding down the whole. This is performed only in the spring, and has the advantage of being so contrived that the stranger bud is placed immediately below that part of a branch where the processes of organisation are most active, namely, below a central bud of the stock; and, from occupying all the circumference, it must necessarily receive the whole of the alimentary and organised matter sent downwards by that bud. It is employed in Bavaria for the Mulberry. (See *Gard. Mag.*, v. 425.)

In GRAFTING no attempt is made to apply the inner surface of the bark of a scion to the outer surface of the wood of the stock ; but the contact is effected by the wood of the two, and their bark only joins at the edges. Whip-grafting (*fig. 27.*) is the commonest kind ; it is performed by heading down a stock, then paring one side of it bare for the space of an inch or so, and cutting down obliquely at the upper end of the pared part, towards the pith ; the scion is levelled obliquely to a length corresponding with the pared surface of the stock, and an incision is made into it near the upper end of the wound obliquely upwards, so as to form a

27 "tongue," which is forced into the corresponding wound in the stock; care is then taken that the bark of the scion is exactly adjusted to that of the stock, and the two are bound firmly together.

Here the mere contact of the two enables the sap flowing upwards through the stock to sustain the life of the scion until the latter can develope its buds, which then send downwards their wood; at the same time the cellular system of the parts in contact unites by granulations; and, when the wood descends, it passes through the cellular deposit, and holds the whole together. The use of "tongueing" is merely to steady the scion, and to prevent its slipping. The advantage of this mode of grafting is, the quickness with which it may be performed; the disadvantage is, that the surfaces applied to each other are much smaller than can be secured by other means. It is, however, a great improvement upon the old crown-grafting, still employed in the rude unskilful practice of some Continental gardeners, but expelled from Great Britain; which consists of nothing more than heading down a stock with an exactly horizontal cut, and splitting it through the middle, into which is forced the end of a scion cut into the form of a wedge; when the whole are bound together. In this method the

split in the stock can hardly be made to heal without great care ; the union between the edges of the
scion and those of the stock is very imperfect, because the bark of the former necessarily lies upon
the wood of the latter, except just at the sides; and,

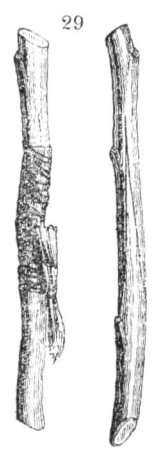

from the impossibility of bringing the two
barks in contact, neither the ascending nor
descending currents of sap are able freely
to intermingle. This plan, much improved
by cutting out the stock into the form of a
wedge, instead of splitting it, may, however, be advantageously employed for such plants
as Cactaceæ (*fig.* 28.), the parts of which, owing
to their succulence, readily form a union with
each other.

A far better method than whip-
grafting, but more tedious, is saddle-
grafting (*fig.* 29.) ; in which the
stock is pared obliquely on both sides
till it becomes an inverted wedge,
and the scion is slit up the centre,
when its sides are pared down till
they fit the sides of the stock.   In
this method, the greatest possible
quantity of surface is brought into
contact, and the parts are mutually
so adjusted, that the ascending sap
is freely received from the stock by the scion, while,
at the same time, the descending sap can flow freely

from the scion into the stock. Mr. Knight, in describing this mode of operating, has the following observations : —

" The graft first begins its efforts to unite itself to the stock just at the period when the formation of a new internal layer of bark commences in the spring ; and the fluid which generates this layer of bark, and which also feeds the inserted graft, radiates in every direction from the vicinity of the medulla to the external surface of the alburnum. The graft is, of course, most advantageously placed when it presents the largest surface to receive such fluid, and when the fluid itself is made to deviate least from its natural course. This takes place most efficiently when (as in this saddle-grafting) a graft of nearly equal size with the stock is divided at its base and made to stand astride the stock, and when the two divisions of the graft are pared extremely thin, at and near their lower extremities, so that they may be brought into close contact with the stock (from which but little bark or wood should be pared off) by the ligature." (*Hort. Trans.*, v. 147.) To execute saddle-grafting properly, the scion and stock should be of equal size ; and where that cannot be, a second method, in which the scion may be much smaller than the stock, has been described by the same great gardener. This (*fig.* 30.) is practised upon small stocks almost exclusively in

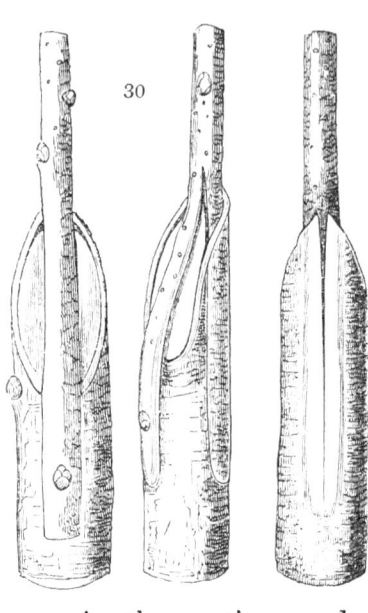

30

Herefordshire ; but it is never attempted till the usual season of grafting is past, and till the bark is readily detached from the alburnum. The head of the stock is then taken off, by a single stroke of the knife, obliquely, so that the incision commences about the width of the diameter of the stock below the point where the medulla appears in the section, and ends as much above it, upon the opposite side. The scion, or graft, which should not exceed in diameter half that of the stock, is then to be divided longitudinally, about two inches upwards from its lower end, into two unequal divisions, by passing the knife upwards, just in contact with one side of the medulla. The stronger division of the graft is then to be pared thin at its lower extremity, and introduced, as in crown-grafting, between the bark and wood of the stock ; and the more slender division is fitted to the stock upon the opposite side. The graft, consequently, stands astride the stock, to which it attaches itself firmly upon each side, and which it covers completely in a single season. Grafts of the Apple and Pear rarely ever fail in this method of grafting, which may be practised with equal success with young wood in July, as soon as it has become moderately firm and mature.

In all these methods, and in every other that could be named, it is indispensable that similar parts should be brought as much as possible into contact; for the more completely this is accomplished, the more certain is the operation to succeed. It is undoubtedly true, that, as the cellular system of a tree is diffused through its whole diameter (43. 46.), it is impossible to apply a scion to a stock without their cellular systems coming in contact; and, therefore, it might appear indifferent whether bark is applied to bark and alburnum to alburnum, or whether the bark is adapted to the wood and the latter to the liber. But it is always to be remembered that each of these parts has special modifications of its own, which modifications require contact with parts similarly modified, in order to unite readily and firmly; and also, that, although the cellular horizontal system, through which union by the first intention takes place, may be alive on all parts of the section of a branch, yet that it is in the bark, and in the space between the bark and wood, that its developement is most rapid, and its tendency to growth most easily excited and maintained.

It is not, however, to be supposed that these operations can be performed indifferently between any two species, although such was formerly so general a belief that it was asserted that roses became black when grafted on Black Currants, and oranges crimson if worked on the Pomegranate.* In point of fact, the operations are successful in those cases only where the stock and scion are very nearly allied; and the degree of success is in proportion to the degree of affinity. Thus, varieties of the same species unite the most freely, then species of the same genus, then genera of the same natural order; beyond which the power does not extend, unless, in the case of parasites like the Mistletoe, which grow indifferently upon totally different plants. For instance, Pears work freely upon Pears, very well on Quinces, less willingly on Apples or Thorns, and not at all upon Plums or Cherries; while the Lilac will take on the Ash, and the Olive on the Phillyrea, because they are plants of the same natural order. M. DeCandolle even says that he has succeeded, notwithstanding the great difference in their vegetation, to work the Lilac on the Phillyrea, the Olive on the Ash, and the Bignonia radicans on

---

* Et steriles platani malos gessêre, valentes
  Castaneæ fagos, ornusque incanuit albo
  Flore pvri, glandemque sues fregêre sub ulmis.

  Georg. lib. ii.

the Catalpa ; but plants so obtained are very short-lived. For some curious particulars upon this subject, see *Physiologie Végétale*, p. 788., &c.

There are two cases apparently at variance with this law ; both of which require explanation.

1. Columella asserts that, by a particular manner of grafting, the Olive may be made to take upon the Fig tree, and his words have been repeated by many writers ; but Thouin proved, experimentally, that no such union will take place, and that where success appears to attend Columella's operation, it is owing to the scion rooting into the soil, independently of the Fig stock (see *Mémoire sur la prétendue Greffe Columelle*), and becoming a layer.

2. Mention is made by Pliny of a tree in the garden of Lucullus, which was so grafted as to bear pears, apples, figs, plums, olives, almonds, grapes, &c. ; and at this day the gardeners of Italy, especially of Genoa and Florence, sell plants of Jasmines, Roses, Honeysuckles, &c., all growing together from a stock of Orange, or Myrtle, or Pomegranate, on which they say they are grafted. But this is a mere cheat, the fact being that the stock has its centre bored out, so as to be made into a hollow cylinder, through which the stems of Jasmines and other flexible plants are easily made to pass, their roots intermingling with those of the stock ; after growing for a time, the horizontal distension of the stems forces them together, and

they assume all the appearances of being united. Such plants are, of course, very short lived.

From what has been now stated, it may be easily conceived that the choice of the stock on which a given plant is to be worked is by no means a matter of indifference, but that the operation may be seriously affected by the skill with which the most suitable stock is selected.   If, indeed, we had no other object in view in grafting than to unite one plant to another, that object would doubtless be best attained by using the same species, and even a similar variety of the same species, for both stock and scion ; the ends of grafting and budding are, however, much beyond this, and it often happens that the species to which a scion belongs, or the nearest variety, is the worst on which it can be worked.   It is, indeed, sometimes asserted that the stock exercises little influence over the scion, but this is so great an error that it cannot be too distinctly contradicted.   This subject has already been adverted to, but it now requires more special consideration.

One of the first objects of budding and grafting is, to multiply a given species or variety more readily than is possible by any other method.   If this is the only purpose of the cultivator, that stock will obviously be the best which can be most readily procured ; and hence we see, in the ordinary practice of the nurseries, the common Plum

taken as a stock for Peaches and Apricots, the
Wild Pear and Crab for Pears and Apples, and so
on.   When there is a difficulty in procuring a
suitable stock, pieces of the roots of the plant to
be multiplied are often taken as a substitute, and
they answer the purpose perfectly well; for the
circumstance which hinders the growth of pieces
of a root into young branches is merely their want
of buds : if a scion is grafted upon a root, that
deficiency is supplied, and the difference between
the internal organisation of a root and a branch is
so trifling as to oppose no obstacle to the solid
union of the two.

Mr. Knight was the first to show the possibility
of grafting scions upon roots.   An account of his
method of doing this was given at a very early
period of the existence of the Horticultural Society (June, 1811), and he at the same time sug-
gested the possibility of the practice being applied
to grafting scarce herbaceous plants upon the
roots of their commoner congeners ; an operation
now commonly practised with the Dahlia, Pæony,
and other plants of a similar kind ; and lately a
method of multiplying Combretum purpureum by
similar means has been pointed out in the *Proceedings of the Horticultural Society*, i. 40.

Mere propagation is, however, by no means the
only object of the grafter ; another and still more
important one is, to secure a permanent union be-

tween the scion and stock, so that the new plant
may grow as freely and as long as if it were on
its own bottom under the most favourable circumstances.   If this is not attended to, the hopes
of the cultivator will be frustrated by the early
death of his plant.

Whenever the stock and graft or bud are not
perfectly well suited to each other, an enlargement
is well known always to take place at the point of
their junction, and generally to some extent either
above or below it.   This is particularly observable in Peach trees which have been budded, at
any considerable height from the ground, upon
Plum stocks ;   and it appears to arise from the
obstruction which the descending sap of the Peach
tree meets with in the bark of the Plum stock ;
for the effects produced, both upon the growth and
produce of the tree, are similar to those which
occur when the descent of the sap is impeded by a
ligature, or by the destruction of a circle of bark.
In course of time this difference between the scion
and stock puts an end to the possibility of the
ascending and descending fluids passing into each
other, and the death of the scion is the result.   In
all the cases I have seen, this has arisen from the
power of horizontal growth in the stock and scion
being different ; and I doubt whether it ever proceeds from any other cause.   For example : the
Hawthorn and the Pear are so nearly allied that

the latter may be easily worked upon the former; the Hawthorn is, however, a slow-growing bush or small tree, the Pear is a large forest tree of rapid growth; and the Pear will grow an inch in diameter while the Hawthorn is growing half an inch.

This last circumstance, if the difference in the rate of growth or in other respects is not excessive, may be taken advantage of for particular purposes. When trees grow too large for a small garden, it is desirable to dwarf them; and when they are naturally unfruitful, to render them productive; both which effects result, at the same time, from grafting them upon stocks that grow slower than themselves. Thus the Apple is dwarfed by grafting on the Paradise stock, and the Pear by the Quince. The physiological explanation of trees dwarfed by being compelled to grow upon a stock which compels their descending sap to accumulate in the branches has been already given (85.). Instead of repeating it here, I take the following paragraph from the paper by Mr. Knight, " On the Effects of different Kinds of Stocks in Grafting," published in the *Horticultural Transactions*, ii. 199.

" The disposition in young trees to produce and nourish blossom buds and fruit is increased by this apparent obstruction of the descending sap; and the fruit of such young trees ripens, I think, somewhat earlier than upon other young trees of

the same age, which grow upon stocks of their
own species ; but the growth and vigour of the
tree, and its power to nourish a succession of heavy
crops, are diminished, apparently, by the stagnation, in the branches and stock, of a portion of that
sap which, in a tree growing upon its own stem,
or upon a stock of its own species, would descend
to nourish and promote the extension of the roots.
The practice, therefore, of grafting the Pear tree
on the Quince stock, and the Peach and Apricot
on the Plum, where extensive growth and durability are wanted, is wrong ; but it is eligible where-
ever it is wished to diminish the vigour and growth
of the tree, and where its durability is not thought
important.

" When," adds this great gardener, " much difficulty is found in making a tree, whether fruc-
tiferous or ornamental, of any species or variety,
produce blossoms, or in making its blossoms set
when produced, success will probably be obtained
in almost all cases by budding or grafting on a
stock which are nearly enough allied to the graft to
preserve it alive for a few years, but not permanently. The Pear tree affords a stock of this kind to
the Apple ; and I have obtained a heavy crop of
Apples from a graft which had been inserted in a
tall Pear stock only twenty months previously, in
a season when every blossom of the same variety
of fruit in the orchard was destroyed by frost.

The fruit thus obtained was externally perfect, and possessed all its ordinary qualities; but the cores were black and without a single seed; and every blossom had certainly fallen abortively, if it had been growing upon its native stock. The experienced gardener will readily anticipate the fate of the graft; it perished in the following winter. The stock, in such cases as the preceding, promotes, in proportion to its length, the early bearing and early death of the graft."

It is sometimes desirable to increase the hardiness of a variety, and grafting or budding appears to produce this effect to a certain extent, not, indeed, by the stock communicating to the scion any of its own power of resisting cold, but by the stock being better suited to the soil of latitudes colder than that from which the scion comes, and consequently requiring a lower bottom-heat to arouse its excitability. Mr. Knight, indeed, denies this fact, because "the root which nature gives to each seedling plant must be well, if not best, calculated to support it;" and it is so, under the circumstances in which the species was first created; but, without this addition, the paragraph quoted in inverted commas is specious only, not just. Probably, in Persia, the native country of the Peach, that species, or its wild type the Almond, is the best stock for the former fruit; because the temperature of the earth (see 117, 118, 119., and

Book II. Ch. I.) is that in which it was created to grow. But in a climate like that of England, the temperature of whose soil is so much lower than that of Persia, the Plum, on which the Peach takes freely, is a hardy native, and suited to such soil, and its roots are aroused from their winter sleep by an amount of warmth unsuited to the Peach. And experience, in this case, completely confirms what theory teaches : for, although there may be a few healthy trees in this country growing upon Almond stocks, it is perfectly certain that the greater part of those which have been planted have failed; while, in the warm soil of France and Italy, it is the stock upon which the old trees have, in almost all cases, been budded.

In determining upon what kind of stock a given fruit tree should be grafted, it is important to be aware that certain species prefer particular soils and dislike others, for reasons which are not susceptible of explanation. In the case of the common stocks employed for the propagation of the Apple, Pear, Peach, and Cherry, it was found by Mr. Dubreuil, an intelligent gardener at Rouen, that in the chalky gardens about that city neither the Plum nor the wild Cherry would succeed for stone fruit, nor the Doucin or Quince stock for Pears and Apples; but that the Crab suited the Apple, the wild Pear the cultivated Pear, the Almond the Peach, and the Mahaleb the Cherry.

I formerly witnessed the result of those experiments while in progress, and I well remember the sickly state of his Peaches and Cherries grafted on Plum and Cherry stocks in the calcareous borders of the rampart gardens of Rouen, and the healthiness of the same fruit trees in the same garden, when worked upon the Almond and the Mahaleb, while the latter were unhealthy in their turn in the borders composed artificially of loam. The result of this experiment has been mentioned in the *Hort. Trans.*, iv. 566., and is as follows: —

	Loamy Soil.	Chalky.	Light.*
Apple - -	Doucin	Crab	Doucin.
Pear - -	Quince	Wild Pear	Quince.
Plum - -	Plum	Almond	Almond.
Cherry - -	Wild Cherry	Mahaleb	Wild Cherry.

As this work treats exclusively of those operations in gardening which can be explained upon known principles of vegetable physiology, all further reference to the question of stocks ought, in strictness, to be dismissed at this stage. It may be as well, however, to add that there are some well attested facts relating to the preference of particular varieties for one kind of stock rather than another, which we cannot explain, but which are so important in practice as to deserve to be studied carefully. There appears to be no doubt that, as is asserted by Mr. Knight and others (*Hort.*

* That is, with an admixture of sand and decayed vegetable matter.

Trans., ii. 215.; *Gard. Mag.*, vii. 195.), the Apricot succeeds better on its own species than on the Plum. The nurserymen know very well that what they call French Peaches, such as the Bourdine, Belle Chevreuse, and Double Montagne, will only take on the Pear Plum, while other varieties prefer the Muscle Plum ; and a variety called the Brompton suits them all equally well, making handsome trees, which are, however, uniformly short-lived.* The Lemon is also found to be a better stock for the Orange than its own varieties.

It is not merely upon the productiveness or vigour of the scion that the stock exercises an influence ; its effects have been found to extend to the quality of the fruit. This may be conceived to happen in two ways — either by the ascending sap carrying up with it into the scion a part of the secretions of the stock, or by the difference induced in the general health of a scion by the manner in which the flow of ascending and descending sap is promoted or retarded by the stock. In the Pear, the fruit becomes lighter coloured, and smaller on the Quince stock, than on the wild Pear, still more so on the Medlar ; and in these two instances the ascent and descent of sap is obstructed by the Quince more than by the wild Pear, and by the Medlar more than by the Quince.

* See G. Lindley's *Guide to the Orchard and Kitchen Garden*, p. 299.

Similar effects are produced in the Apple by the Paradise and Siberian Bittersweet stocks. Mr. Knight mentions such differences in the quality of his Peaches. His garden contained two trees of the Acton Scott variety, " one growing upon its native stock, the other upon a Plum stock, the soil being similar, and the aspect the same. That growing upon the Plum stock afforded fruit of a larger size, and its colour, where it was exposed to the sun, was much more red ; but its pulp was more coarse, and its taste and flavour so inferior that he would have denied the identity of the variety, had he not with his own hand inserted the buds from which both sprang. (*Hort. Trans.*, v. 289.)

In addition to a judicious adaptation of the bud or scion to the stock, there are other circumstances to which it is necessary to attend, in order to insure the success of the operation. It has already been seen (p. 196.), that the youngest buds of the Potato are more excitable than those more completely matured ; and the same appears to be true of the buds in other fruits.

" The mature bud," says Mr. Knight, " takes immediately with more certainty, under the same external circumstances : it is much less liable to perish during winter ; and it possesses the valuable property of rarely or never vegetating prematurely in the summer, though it be inserted before the usual period, and in the season when the sap of the stock is most abundant. I have, in different years, removed some hundred buds of the Peach tree from the forcing-house to luxuriant shoots upon the open wall; and I have never seen an instance in which any of such buds have broken and vegetated during the summer and autumn; but when I have had occasion to reverse this process and to insert immature buds from the open wall into the branches of trees growing in a Peach-house, many of these, and in some seasons all, have broken soon after being inserted, though at the period of their insertion the trees in the Peach-house had nearly ceased to grow." (*Hort. Trans.*, iii. 136.)

This property was turned to practical account by Mr. Knight in budding the Walnut. Owing to the excitability of its buds, this tree is difficult to work, because its buds exhaust all their organisable and alimentary matter before any adhesion can be formed between themselves and the stock; but by taking the small, fully matured, and little developed buds, found at the base of the annual shoots of this plant, time is given for an adhesion between them and the alburnum before they push forth, and then they *take* freely enough. (See *Hort. Trans.*, iii. 135.)

Buds should either be inserted when the vegetation of a plant is languid, or growth above the place of insertion should be arrested by pinching the terminal bud; otherwise the sap, which should be directed into the bud, in order to assist in its adhesion, is conveyed to other places, and the bud perishes from starvation. For similar reasons, when a bud begins to grow, having firmly fixed itself upon the stock, the latter should be headed back nearly as far as the bud, so as to compel all the ascending current of sap to flow towards it; otherwise the buds of the stock itself will obtain that food which the stranger bud should be supplied with.

In grafting also it is always found that a union between the scion and the stock takes place most readily when the latter is headed down; but this is not the only point to attend to. The scion should always be so prepared that a bud is near the point of union between itself and the stock; because such a bud, as soon as it begins to grow, proceeds to furnish wood, which assists in binding the two together. The scion should be more backward in its vegetation than the stock, because it will then be less excitable; otherwise its buds may begin to grow before a fitting communication is established between the stock and scion, and the latter will be exhausted by its own vigour: if, on the contrary, the stock is in the state of incipient growth, and the scion torpid, granulations of cellular tissue will have time to form and unite the wound, and the scion will become distended with sap forced into it from the stock, and thus be able to keep its buds alive when they begin to shoot into branches. In order to assist in this part of the operation, a "heel" is some-times in difficult cases left on a scion, and inserted into a vessel of water, until the union has taken place; or, for the same purpose, the scion is bound round with loose string or linen with one end steeped in water, so as to secure a supply of water to the scion by the capillary attraction of such a band-age. Indeed, the ordinary practice of surrounding the scion and stock at the point of contact with a mass of grafting clay is intended for the same purpose; that is to say, to prevent evaporation from the surface of the scion, and to afford a small supply of moisture; and hence, among other things, the superiority of clay over the plasters, mastics, and cements occasionally employed, which simply arrest perspiration, and can never assist in communicating aqueous food to the scion.

Here also must be noticed certain practices, which experience shows to be important, of which theory offers no obvious explanation. Mr. Knight, for example, asserts that cuttings taken from the trunks of seedling old trees grow much more vigorously than those taken from the extremities of bearing branches; and it is an undoubted fact that the Beech, and other trees of a similar kind, cannot be grafted with any success, unless the scions are made of two-years-old wood; one-year-old wood generally fails.

What is called herbaceous grafting, or Tschudy grafting, depends so entirely upon the same principles as common grafting, that a separate notice of it is hardly necessary. Nevertheless, as it is sometimes very useful, a few words may be given to it. When two vigorous branches cross each other, and press together so as not to move, they will often form an organic union; if two apples press together, or if two cucumbers are forced to grow side by side in a space so small as to compel them to touch each other firmly, they also will grow together; and herbaceous grafting is merely an application to practice of this power of soft and cellular parts to unite. In order to secure success, the scion and stock, being pared so as to fit together accurately, are firmly bound to each other, without being crushed; parts in full vegetation, and abounding in sap, are always chosen for the operation, such as the upper parts of annual shoots, near the terminal bud; perspiration is diminished by the removal of some of the leaves of both stock and scion, and by shading (71.); and by degrees, as the union becomes secured, buds and leaves are removed from the stock, in order that all the sap possible may be impelled into the scion. This method, if well managed, succeeds completely in about thirty days, and is useful as a method of multiplying lactescent, resinous, and hard-wooded trees, which refuse to obey more common methods. Baron de Tschudy succeeded in this way in working the Melon on the Bryony (both Cucurbitaceous plants), the Artichoke on the Cardoon (both Cynaras), Tomatoes on Potatoes (both Solanums), and so on.   The following account of managing Coniferæ, where herbaceous grafting is used, is taken from the *Gardener's Magazine*, vol. ii. p. 64., and sufficiently explains the practice : —

" The proper time for grafting pines is when the young shoots have made about three quarters of their length, and are still so herbaceous as to break like a shoot of asparagus.   The shoot of the stock is then broken off about two inches under its terminating bud ; the leaves are stripped off from twenty to twenty-four lines down from the extremity, leaving, however, two pairs of leaves op-

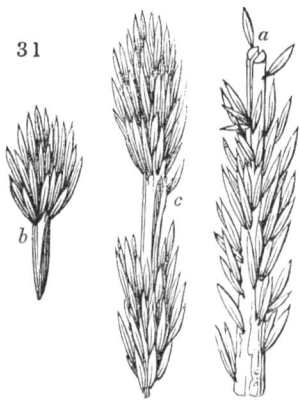

31

posite, and close to the section of fracture, which leaves are of great importance.   The shoot is then split with a very thin knife between the two pairs of leaves (*fig.* 31. *a*), and to the depth of two inches.   The scion is then prepared (*b*) : the lower part, being stripped of its

leaves to the length of two inches, is cut, and inserted in the usual manner of cleft-grafting. They
may also be grafted in the lateral manner (c). The
graft is tied with a slip of woollen, and a cap of paper
is put over the whole, to protect it from the sun and
rain. At the end of fifteen days this cap is removed, and the ligature at the end of a month ;
at that time also the two pairs of leaves (a), which
have served as nurses, are removed. The scions
of those sorts of pines which make two growths in
a season, or, as the technical phrase is, have a
second sap, produce a shoot of five or six inches in
the first year : but those of only one sap, as the
Corsican Pine, Weymouth Pine, &c., merely ripen
the wood grown before grafting, and form a strong
terminating bud, which in the following year produces a shoot of fifteen inches, or two feet."

With regard to INARCHING, which was probably
the most ancient kind of grafting, because it is that
which must take place accidentally in thickets and
forests, it differs from grafting in this, that the scion
is not severed from its parent, but remains attached to it until it has united to the stock to
which it is tied and fitted in various ways ; the
scion and stock are therefore mutually independent
of each other, and the former lives upon its own
resources, until the union is completed.

In practice, a portion of the branch of a scion is
pared away, well down into the alburnum; a cor-

responding wound is made in the branch of a stock; tongues are made in each wound so that they will fit into each other; and the liber and alburnum of the two being very accurately adjusted, the whole are firmly bound up; grafting clay is applied to the wound, and the plants operated upon are carefully shaded ; in course of time the wounds unite, and then the scion is severed from its parent.   Gardeners consider this the most certain of all the modes of grafting, but it is troublesome, and only practised in difficult cases.   The circumstances most conducive to its success are, to stop the branch of both stock and scion under operation, so as to obtain an accumulation of sap, and to arrest the flow of sap upwards; to moderate the motion of the fluids by shading; to head back the stock as far as the origin of the scion, as soon as the union is found to be complete; and at the same time to retrench from the scion a part of its buds and leaves, so that there may not be a a too rapid demand upon the stock, while the line of union is still imperfectly consolidated.

32

A method of propagating Camellias (*fig.* 32.), by putting the end or heel of a scion into a vessel of water, mentioned in the *Gardener's Magazine*, ii. 33., is essentially the same as inarching.

The water communicated to the scion through the wounded end supplies it with that food which, under natural circumstances, would be derived from the roots of the plant to which it belongs.

CHAP. XIII.

OF PRUNING.

" La taille est une des opérations les plus importantes et les plus délicates du jardinage. Confiée communément à des ouvriers peu instruits, observée dans les résultats d'une pratique trop souvent irréfléchie, elle a dû nécessairement trouver des détracteurs même parmi les physiologistes. Il en eût sans doute été autrement, si on l'avait étudiée dans les jardins du petit nombre de praticiens qui ont su de nos jours la bien comprendre. Sagement basée sur les lois de la végétation, elle contribue, entre leurs mains, non seulement à régulariser la production des fruits, à en obtenir de plus beaux, mais encore à prolonger l'existence et la fécondité des arbres."

Nothing can be more just than these words, addressed to the Horticultural Society of Paris, by their President, M. Héricart de Thury ; and, if they do not apply with as much force to our gardeners as to those of France, they do most fully to our foresters.

The quantity of timber that a tree forms, the amount and quality of its secretions, the brilliancy of its colours, the size of its flowers, and, in short, its whole beauty, depend upon the action of its branches and leaves, and their healthiness (64.). The object of the pruner is to diminish the number of leaves and branches; whence it may be at once understood how delicate are the operations he has to practise, and how thorough a knowledge he ought to possess of all the laws which regulate the action of the organs of vegetation. If well directed, pruning is one of the most useful, and, if ill-directed, it is among the most mischievous, operations that can take place upon a plant.

When a portion of a healthy plant is cut off, all that sap which would have been expended in supporting the part removed is directed into the parts which remain, and more especially into those in the immediate vicinity of it. Thus, if the leading bud of a growing branch is stopped, the lateral buds, which would otherwise have been dormant, are made to sprout forth; and, if a growing branch is shortened, then the very lowest buds, which seldom push, are brought into action : hence the necessity, in pruning, of cutting a useless branch clean out ; otherwise the removal of one branch is only the cause of the production of a great many others. This effect of stopping does not always take place immediately ; sometimes its first effect is to cause an accumulation of sap in a branch, which directs itself to the remaining buds, and organises them against a future year. In ordinary cases, it is thus that spurs or short bearing-branches are obtained in great abundance. The growers of the Filbert, in Kent, procure in this way greater quantities of bearing wood than nature unassisted would produce ; for, as the filbert is always borne by the wood of a previous year, it is desirable that every bush should have as much of that wood as can be obtained, for which every thing else may be sacrificed ; and such wood is readily secured by observing a continual system of shortening a young branch by two thirds, the effect of which is to call all its lower buds into growth the succeeding year ; and thus each shoot of bearing wood is compelled to produce many others. The Peach, by a somewhat similar system, has been made to bear fruit in unfavourable climates (*Hort. Trans.*, ii. 366.) ; and every gardener knows how universally it is applied to the Pear, Apple, Plum, and similar trees. Even the Fig-tree has thus been rendered much more fruitful than by any other method. " Whenever," says Mr. Knight, " a branch of this tree appears to be extending with too much luxuriance, its point, at the tenth or twelfth leaf, is pressed between the finger and thumb, without letting the nails come in contact with the bark, till the soft succulent substance is felt to yield to the pressure. Such branch, in consequence, ceases subsequently to elongate ; and the sap is repulsed, to be expended where it is more wanted. A fruit ripens at the base of each leaf, and during the period in which the fruit is ripening, one or more of the lateral buds shoots, and is subsequently subjected to the same treatment, with the same result. When I have suffered such shoots to extend freely to their natural length, I have found that a small part of them only became productive, either in the same or the ensuing season, though I have seen that their buds obviously contained blossoms. I made several experiments to obtain fruit in the following spring from other parts of such branches, which were not successful : but I ultimately found that bending these branches, as far as could be done without danger of breaking them, rendered them extremely fruitful ; and, in the present spring, thirteen figs ripened perfectly upon a branch of this kind within the space of ten inches. In training, the ends of all the shoots have been made, as far as practicable, to point downwards." (*Hort. Trans.*, iv. 201.)

The effect produced upon one part by the abstraction of some other part, thus shown in the developement of buds which would otherwise be dormant, is seen in many other ways. If all the fruit of a plant is abstracted one year when just forming, the fruit will be finer and more abundant the succeeding year, as happens when late frosts destroy our crops. If of many flowers one only is left, that one, fed by the sap intended for the others, becomes so much finer. If the late figs, which never ripen, are abstracted, the early figs the next year are more numerous and larger. If of two unequal branches, the stronger is shortened and stopped in its growth, the other becomes stronger ; and this is one of the most useful facts connected with pruning, because it enables a skilful cultivator to equalise the rate of growth of all parts of a tree ; and, as has been already stated, this is of the greatest consequence in the operation of budding. In fact, the utility of the practice, so common in the management of fruit trees when very young, turns entirely upon this. A seedling tree has a hundred buds to support, and consequently the stem grows slowly, and the plant becomes bushy-headed : but, being cut down so as to leave only two or three buds, they spring upwards with great vigour, and, being reduced eventually to one, as happens practically, that one

receives all the sap, which would otherwise be
diverted into a hundred buds, and thrives accord-
ingly, the bushy head being no longer found, but
a clean straight stem instead.   In the Oak and the
Spanish Chestnut this is particularly conspicuous.

Nothing is more strictly to be guarded against
than the disposition to *bleed*, which occurs in some
plants when pruned, and to such an extent as to
threaten them with death.   In the Vine, in milky
plants, and in most climbers or twiners, this is
particularly conspicuous; and it is not unfrequently
observed in fruit trees with gummy or mucilaginous secretions, such as the Plum, the Peach,
and other stone fruits.   This property usually
arises from the large size of the vessels through
which sap is propelled at the periods of early
growth, which vessels are unable, when cut
through, to collapse sufficiently to close their own
apertures, when they necessarily pour forth their
fluid contents as long as the roots continue to absorb them from the soil.   If this is allowed to
continue, the system becomes so exhausted as to
be unable to recover from the shock, and the plant
will either become very unhealthy, or will die.   The
only mode of avoiding it is to take care never
to wound such trees at the time when their sap
first begins to flow ; after a time, the demand upon
the system by the leaves becomes so great that

there is no surplus, and therefore bleeding does not take place when a wound is inflicted.*

All these things show how extremely necessary it is to perform the operations of pruning with care and discretion. But, in addition to the general facts already mentioned, there are others of a more special kind that require attention. The first thing to be thought of is the peculiar nature of the plant under operation, and the manner in which its special habits may render a special mode of pruning necessary. For example, the fruit of the Fig and Walnut is borne by the wood of the same season; that of the Vine and Filbert by that of the second season; and Pears, Apples, &c., by wood of some years' growth; it is clear that plants of these three kinds will each require a distinct plan of pruning for fruit.

The pruner has frequently no other object in view than that of thinning the branches so as to allow the free access of light and air to the fruit;

* " The Vine often bleeds excessively when pruned in an improper season, or when accidentally wounded ; and, I believe, no mode of stopping the flow of the sap is at present known to gardeners. I therefore mention the following, which I discovered many years ago, and have always practised with success : —If to four parts of scraped cheese be added one part of calcined oyster shells, or other pure calcareous earth, and this composition be pressed strongly into the pores of the wood, the sap will instantly cease to flow ; so that the largest branch may, of course, be taken off at any season with safety." (*Knight,* in *Hort. Trans.,* i. 102.)

and if this purpose is wisely followed, by merely removing superfluous foliage, the end attained is highly useful : it is clear, however, that in order to arrive at this end, without committing injury to the tree which is operated on, it is indispensable that its exact mode of bearing fruit should be in the first instance clearly ascertained.

The period of ripening fruit is sometimes changed by skilful pruning, as in the case of the Raspberry, which may be made to bear a second crop of fruit in the autumn, after the first crop has been gathered. In order to effect this, the strongest canes, which in the ordinary course of things would bear a quantity of fruiting twigs, are cut down to within two or three eyes of the base; the laterals thus produced, being impelled into rapid growth by an exuberance of sap, are unable to form their fruit buds so early as those twigs in which excessive growth is not thus produced, and consequently, while the latter fruit at one season, the others cannot reach a bearing state till some weeks later. Autumnal crops of summer roses, and of strawberries, have been sometimes procured by the destruction of the usual crop at a very early period of the season ; the sap intended to nourish the flower buds destroyed is, after their removal, expended in forming new flower buds, which make their appearance at a later part of the year.

The season for pruning is usually midwinter, or at midsummer; the latter for the purpose of removing new superfluous branches, the former for thinning and arranging the several parts of a tree. It is, however, the practice, occasionally, to perform what is called the winter pruning early in the autumn, as in the case of the Gooseberry, and of the Vine when weak; and the effect is found to be, that the shoots of such plants, in the succeeding season, are stronger than they would have been had the pruning been performed at a much later season. This is necessarily so, as a little reflection will show. During the season of rest (winter) a plant continues to absorb food solely from the earth by its roots (34.); and, if its branches are un-

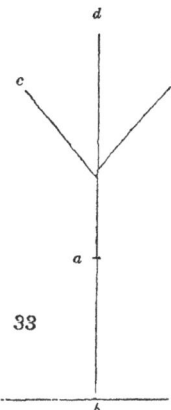

pruned, the sap thus and then introduced into the system will be distributed equally all through it; let us say from *b* to *c d* and *e* in the accompanying diagram. If late pruning is had recourse to, and the branches from *a* to *c d* and *e* are removed, of course a large proportion of the sap that has been accumulating during the winter will be thrown away, and *b* to *c* will retain no more of it than the exact proportion which that part bears to the part abstracted. When, however, early or autumnal pruning is employed, *a* to *c d* and *e* are removed before the sap has accumulated in them, and then all which the roots are capable of collecting during the period of repose will be deposited in the space from *b* to *a*, and consequently branches from that part will necessarily push with excessive vigour. As, however, pruning is by no means intended at all times to increase the vigour of a plant, late or spring pruning, if not deferred till the sap is in rapid motion, may be the more judicious.

With regard to pruning plants when transplanted, there can be no doubt that it is more frequently injurious than beneficial. It is supposed, or seems to be, that when the branches of a transplanted tree are headed back, the remaining buds will break with more force than if the pruning had not been performed ; but it is to be remembered that a transplanted tree is not in the state supposed in the case put at page 259. fig. 33. Its roots are not fully in action, but from the injuries sustained in removing they are capable of exercising but little influence on the branches. The great point to attain, in the first instance, is the renovation of the roots, and that will happen only in proportion to the healthy action of the leaves and buds (31.): if, therefore, the branches of a plant are removed by the pruning-knife, a great obstacle is opposed to this renovation ; but, if they remain, new roots will be formed in proportion to their healthy action.

The danger to be feared is, that the perspiration of the leaves may be so great as to exhaust the system of its fluid contents faster than the roots can restore them, and in careless transplanting this may doubtless happen : in such cases it is certainly requisite that some part of the branches should be pruned away ; but no more should be taken off than the exigency of the case obviously requires ; and, if the operation of transplanting has been well performed, there will be no necessity whatever. In the case of the transplantation of large trees, it is alleged that branches must be removed, in order to reduce the head, so that it may not be acted upon by the wind; but in general it is easy to prevent this action by artificial means.

In the nurseries it is a universal practice to prune the roots of transplanted trees; in gardens this is as seldom performed. Which is right? If a wounded or bruised root is allowed to remain upon a transplanted tree, it is apt to decay, and this disease may spread to neighbouring parts, which would otherwise be healthy ; to remove the wounded parts of roots is therefore desirable. But the case is different with healthy roots. We must remember that every healthy and unmutilated root which is removed is a loss of nutriment to the plant, and that too at a time when it is least able to spare it; and there cannot be any advantage in the removal. The nursery practice is probably intended to render the operation of transplanting large numbers of plants less troublesome; and, as it is chiefly applied to seedlings and young plants with a superabundance of roots, the loss in their case is not so much felt. If performed at all, it should take place in the autumn, for at that time the roots, like the other parts of a plant, are comparatively empty of fluid; but, if deferred till the spring, then the roots are all distended with fluid, which has been collecting in them during winter, and every part taken away carries with it a portion of that nurture which the plant had been laying up as the store upon which to commence its renewed growth.

It must now be obvious that, although root-pruning may be prejudicial in transplanting trees, it may be of the greatest service to such established trees as are too prone to produce branches and leaves, instead of flowers and fruit. In these cases the excessive vigour is at once stopped by removal of some of the stronger roots, and consequently of a part of the superfluous food to which their "rankness" is owing. The operation has been successfully performed on the wall trees at Oulton, by Mr. Errington, one of our best English gardeners, and by many others, and, I believe, has never proved an objectionable practice under judicious management. Its effect is, *pro tanto*, to cut off the supply of food, and thus to arrest the rapid growth of the branches; and the connexion between this and the production of fruit has already been explained (85.). It is by pushing the root-pruning to excess that the Chinese obtain the curious dwarf trees which excite so much curiosity in Europe. Mr. Livingston's account of their practice is so instructive, and contains so much that an intelligent gardener may turn to account, that I think it worth repeating here.

"When the dwarfing process is intended, the branch which had pushed radicles into the surrounding composition in sufficient abundance, and for a sufficient length of time is separated from the tree, and planted in a shallow earthenware flower-pot, of an oblong square shape; it is sometimes made to rest upon a flat stone. The pot is then filled with small pieces of alluvial clay, which, in the neighbourhood of Canton, is broken into bits, of about the size of common beans, being just sufficient to supply the scanty nourishment which the particular nature of the tree and the process require. In addition to a careful regulation of the quantity and quality of the earth, the quantity of water, and the management of the plants with respect to sun and shade, recourse is had to a great variety of mechanical contrivances, to produce the desired shape. The con-

s 4

taining flower-pot is so narrow, that the roots
pushing out towards the sides are pretty effectually cramped.   No radicle can descend;
consequently it is only those which run towards
the sides or upwards that can serve to convey
nourishment properly, and it is easy to regulate those by cutting, burning, &c., so as to cramp
the growth at pleasure.   Every succeeding formation of leaves becomes more and more stunted,
the buds and radicles become diminished in the
same proportion, till at length that balance between the roots and leaves is obtained which suits
the character of the dwarf required.   In some
trees this is accomplished in two or three years,
but in others it requires at least twenty years."
(*Hort. Trans.*, iv. 229.)

We have still to consider that peculiar kind of
pruning which is technically called *ringing* (*fig.* 34.).
This consists in removing from a branch one or
more rings of bark, by which the return of sap
from the extremities is obstructed, and it is compelled to accumulate above the ring.   Mr. Knight
explains the physiological nature of the operation
so well, that I cannot do better than quote his words.

" The true sap of trees is wholly generated in
their leaves, from which it descends through their
bark to the extremities of their roots, depositing
in its course the matter which is successively added
to the tree; whilst whatever portion of such sap

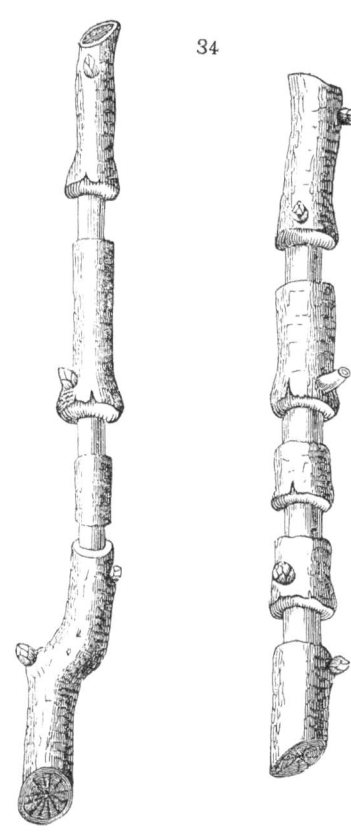

34

is not thus expended
sinks into the alburnum,
and joins the ascending
current, to which it com-
municates powers not
possessed by the re-
cently absorbed fluid.
When the course of the
descending current is
intercepted, that natu-
rally stagnates and accu-
mulates above the de-
corticated space; whence
it is repulsed and carried
upwards, to be expended
in an increased produc-
tion of blossoms, and of
fruit : and, consistently
with these conclusions, I
have found that part of
the alburnum which is
situated above the decorticated space to exceed
in specific gravity, very considerably, that which
lies below it. The repulsion of the descending
fluid, therefore, accounts, I conceive, satisfactorily,
for the increased production of blossoms, and more
rapid growth of the fruit upon the decorticated
branch : but there are causes which operate in promoting its more early maturity. The part of the

branch which is below the decorticated space is ill
supplied with nutriment, and ceases almost to
grow; it in consequence operates less actively in
impelling the ascending current of sap, which must
also be impeded in its progress through the decorticated space. The parts which are above it must,
therefore, be less abundantly supplied with moisture, and drought, in such cases, always operates
very powerfully in accelerating maturity. When
the branch is small, or the space from which the
bark has been taken off is considerable, it almost
always operates in excess; a morbid state of early
maturity is induced, and the fruit is worthless.

" If this view. of the effects of partial decorti-
cation, or ringing, be a just one, it follows that
much of the success of the operation must be dependent upon the selection of proper seasons, and
upon the mode of performing it being well adapted
to the object of the operator. If that be the production of blossoms, or the means of making the
blossoms set more freely, the ring of bark should
be taken off early in the summer preceding the
period at which blossoms are required : but, if the
enlargement and more early maturity of the fruit
be the objects, the operation should be delayed till
the bark will readily part from the alburnum in the
spring. The breadth of the decorticated space
must be adapted to the size of the branch ; but I
have never witnessed any except injurious effects,

whenever the experiment has been made upon
very small or very young branches, for such become debilitated and sickly, long before the fruit
can acquire a proper state of maturity."

The effects of ringing in altering the appearance
of the fruit is very striking. In the *Horticultural
Transactions*, iii. 367., the following cases are re-
ported : — In a French Crab, the fruit, by ringing,
was increased to more than double the size, and the
colour of it was much brightened. In a Minshull
Crab the size was not increased, but the appearance of the apple was so improved as to make it
truly beautiful ; its colours, both red and yellow,
were very bright. In the Court-pendu Apple the
improvement was still more conspicuous, the colours being changed from green and dull red, to
brilliant· yellow and scarlet. Many others of a
similar kind are to be found recorded in books on
horticulture. It is, however, by no means alone to
the maturation or production of fruit that this operation is applicable ; it will, of course, induce also
the production of flowers, and it has occasionally
been used for that purpose, as in the Camellia.
It is best performed in the early spring, when the
bark first separates freely from the wood.

This operation has, however, the disadvantage
of wounding a branch severely ; and, if performed
extensively upon a tree, it is very apt, if not to
kill it, at least to render it incurably unhealthy ;

for if the rings are not sufficiently wide to cut off
all communication between the upper and lower
lips of the wound they produce little effect, and if
they are, they are difficult to heal.   For these reasons the operation is but little employed, other
means being used instead. By some persons ligatures
are made use of, and they would be preferable if
they answered the purpose of obstructing the sap
to the same extent as the abstraction of a ring of
bark.   In Malta, one of the objects of ringing,
that of advancing the maturation of the fruit, is
practised upon the Zinzibey, or Jujube tree, by
merely fixing in the fork of a branch a very heavy
stone, made fast with bandages; its weight forces
the branches a little into a horizontal direction,
and thus, independently of the pressure it exercises upon the parts it touches, obstructs the free
circulation of the sap.

CHAP. XIV.

OF TRAINING.

TRAINING is one of the most artificial operations
that gardeners are acquainted with, its object being
to place a plant in a condition to which it could

never arrive under ordinary circumstances. The practice of it forms one of the most complicated parts of the art of horticulture, each species demanding a method peculiar to itself; but the principles on which it depends are few and simple. These will be best considered with reference to the objects the gardener wishes to attain in performing the operation.

It is probable that the intention of the first gardener who trained a tree was to gain some advantage of climate, by placing the tree close to a wall or other screen; and this is still one of the greatest objects; partly with a view to guard the flowers in spring from cold, and especially cold winds, partly to expose the leaves and fruit to a hotter temperature than would otherwise be gained, and in some measure to ripen wood with more certainty.

That training a tree over the face of a wall will protect the blossoms from cold must be apparent, when we consider the severe effect of excessive evaporation upon the tender parts; a merely low temperature will produce but little comparative injury in a still air, because the more essential parts of the flower are very much guarded by the bracts, calyx, and petals, which overlie them, and, moreover, because radiation (see page 138.) will be intercepted by the branches themselves placed one above the other, so that none but the uppermost branches which radiate into space will feel its full effects ; but, when a cold wind is constantly passing through the branches and among the flowers, the perspiration, against which no sufficient guard is provided by nature, becomes so rapid (see page 132.) as to increase the amount of cold considerably, besides abstracting more aqueous matter than a plant can safely part with. This being one of the great objects of training trees, it is inconceivable how any one should have recommended such devices as those mentioned in the *Horticultural Transactions*, ii. Appendix, p. 8., of training trees upon a horizontal plane ; the only effect of which would be to expose a tree as much as possible to the effect of that radiation which it is the very purpose of training to guard against.

The actual temperature to which a tree trained upon a wall facing the sun is exposed is much higher than that of the surrounding air, not only because it receives a larger amount of the direct solar rays, but because of the heat received by the surrounding earth, reflected from it and absorbed by the wall itself. Under such circumstances the secretions of the plant are more fully elaborated than in a more shady and colder situation; and, by aid of the greater heat and dryness in front of a south wall, the period of maturity is much advanced. In this way we succeed in procuring a Mediterranean or Persian summer in these northern latitudes. When the excellence of fruit depends upon its sweetness, the quality is exceedingly improved by such an exposure to the sun; for it is found that the quantity of sugar elaborated in a fruit is obtained by an alteration of the gummy, mucilaginous, and gelatinous matters previously formed in it, and the quantity of those matters will be in proportion to the amount of light to which the tree, if healthy, has been exposed. Hence the greater sweetness of plums, pears, &c., raised on walls from those grown on standards. It has been already stated (page 141.) that an increase of heat has been sought for on walls by blackening them; and we are assured in the *Horticultural Transactions* (iii. 330.) that, in the cultivation of the Grape, this has been attended with the best effects. But, unless when trees are young, the wall ought to be covered with foliage during summer, and the blackened surface would scarcely act; and in the spring the expansion of the flowers would be hastened by it, which is no advantage in cold late springs, because of the greater liability of early flowers to perish from cold. That a blackened surface does produce a beneficial effect upon trees trained over it is, however, probable, although not in insuring the maturation of fruit; it is by raising the temperature of the wall in autumn when the leaves are falling, and the darkened surface becomes uncovered, that the advantages are perceived by a better completion of the process of growth, the result of which is the ripening the wood. This is, indeed, the view taken of it by Mr. Harrison, who found the practice necessary, in order to obtain crops of pears in late seasons at Wortley in Yorkshire (see *Hort. Trans.*, iii. 330. and vi. 453.). It hardly need be added that the effect of blackening will be in proportion to the thinness of the training, and *vice versâ.*

Another object of training is, to place a tree in such a state of constraint that its juices are unable to circulate freely, the result of which is exactly that already assigned to the process of ringing (see p. 267.). If a stem is trained erect it will be more vigorous than if placed in any other position, and its tendency to bear leaves rather than flowers will be increased ; in proportion as it deviates from the perpendicular is its vigour diminished. For instance, if a stem is headed back, and only two opposite buds are allowed to grow, they will continue to push equally, so long as their relation to the perpendicular is the same ; but, if one is bent towards a horizontal direction, and the other allowed to remain, the growth of the former will be immediately checked; if the depression is increased, the weakness of the branch increases proportionally ; and this may be carried on till the branch perishes. In training, this fact is of the utmost value in enabling the gardener to regulate the symmetry of a tree. It however by no means follows that, because out of two contiguous branches, one growing erect and the other forced into a downward direction, the latter may die, that all branches trained downwards will die. On the contrary, an inversion of their natural position is of so little consequence to their healthiness, that no effect seems in general to be produced beyond that of causing a slow circulation, and the formation of flowers. Hence the directing of branches downwards is one of the commonest and most successful contrivances employed by gardeners to render plants fruitful. Mr. Knight was the first to recommend the practice, in the following account of his recovery of an old and worthless Pear tree.

" An old St. Germain Pear tree, of the spurious kind, had been trained in the fan form, against a north-west wall in my garden, and the central branches, as usually happens in old trees thus trained, had long reached the top of the wall, and had become wholly unproductive. The other branches afforded but very little fruit, and that never acquiring maturity was consequently of no value ; so that it was necessary to change the variety, as well as to render the tree productive. To attain these purposes, every branch which did not want at least twenty degrees of being perpendicular was taken out at its base ; and the spurs upon every other branch, which I intended to re-

T

tain, were taken off closely with the saw and chisel. Into these branches, at their subdivisions, grafts were inserted at different distances from the root, and some so near the extremities of the branches, that the tree extended as widely in the autumn after it was grafted, as it did in the preceding year. The grafts were also so disposed, that every part of the space the tree previously covered was equally well supplied with young wood.

" As soon, in the succeeding summer, as the young shoots had attained sufficient length, they were trained almost perpendicularly downwards, between the larger branches and the wall, to which they were nailed. The most perpendicular remaining branch upon each side was grafted about four feet below the top of the wall, which is twelve feet high ; and the young shoots, which the grafts upon these afforded, were trained inwards, and bent down to occupy the space from which the old central branches had been taken away ; and therefore very little vacant space remained any where in the end of the first autumn. A few blossoms, but not any fruit, were produced by several of the grafts in the succeeding spring ; but in the following year, and subsequently, I have had abundant crops, equally dispersed over every part of the tree ; and I have scarcely ever seen such an exuberance of blossom as this tree presents in the present spring." (*Hort. Trans.*, ii. 78.)

The practice was then followed by Sir Joseph

Banks, whose fruit trees trained downwards over the walls of his garden at Spring Grove, and facing the high road, long excited the astonishment of passers by ; and it has now been generally applied to other cases. What are called Balloon Apples and Pears, formed by forcing downwards all the branches of standard trees till the points touch the earth, are an instance of this ; and they have the merit of producing large crops of fruit in a very small compass : their upper parts are, however, too much exposed to radiation at night, and the crop from that part of the branches is apt to be cut off. One of the prettiest applications of this

35

principle is that of Mr. Charles Lawrence, described in the *Gardener's Magazine*, viii. 680., by
means of which standard Rose trees are converted
into masses of flowers. The figure given in that
work, and here reproduced (*fig.* 35.), represents
the variety called the Bizarre de la Chine, "which
flowered most abundantly to the ends of its branches,
and was truly a splendid object."

The last object of training to which it is necessary to advert is that of improving the quality of
fruit, by compelling the sap to travel to a very
considerable distance. The earliest notice of this,
with which I am acquainted, is the following by
Mr. Williams of Pitmaston.

"Within a few years past," he says in 1818,
"I have gradually trained bearing branches of a
small Black Cluster Grape, to the distance of near
fifty feet from the root, and I find the branches
every year grow larger, and ripen earlier as the
shoots continue to advance. According to Mr.
Knight's theory of the circulation of the sap, the
ascending sap must necessarily become enriched by
the nutritious particles it meets with in its progress
through the vessels of the alburnum ; the wood at
the top of tall trees, therefore, becomes short-
jointed and full of blossom buds, and the fruit
there situated attains its greatest perfection. Hence
we find Pine and Fir trees loaded with the finest
cones on the top boughs ; the largest acorns grow
on the terminal branches of the Oak, and the

finest mast on the high boughs of the Beech and
Chestnut; so likewise apples, pears, cherries, &c.,
are always best flavoured from the top of the tree."
(*Hort. Trans.*, iii. 250, 251.) The merit of the
Fontainbleau mode of training the Vine (*fig.* 36.),
in which many of the stems are carried to very

36

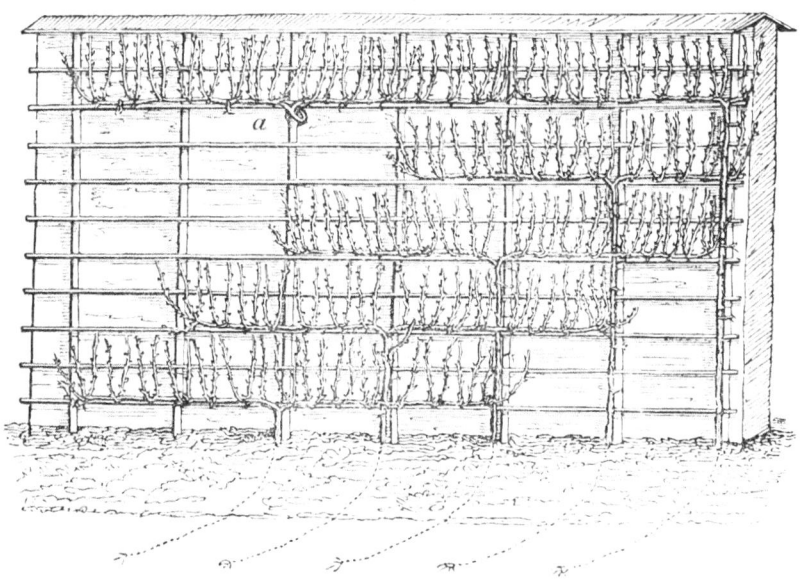

considerable distances, seems to depend in some
measure upon this principle ; and there is a well-
known Black Hamburg Grape at Bath, growing
in a garden formerly belonging to Mr. Farrant, the
stem of which, owing to local circumstances, is
necessarily conveyed to a very considerable distance before it is allowed to produce its bearing

branches, the quality of whose fruit is of very unusual excellence. These facts seem capable of being applied to many important improvements in fruit management.

The foregoing are the principal advantages which arise from training plants; let us next consider what disadvantages there may be. The only trees which at all approach in nature the state of trained plants are climbers and creepers, whose stems, unable to support themselves, cling for a prop upon whatever they are near ; some of them enclose the stem of another plant in their convolutions; others simply attach themselves by means of tendrils as the Vine, by hooks as the Combretum, or by other contrivances ; and some, like the Ivy, lay hold of walls, rocks, or the trunks of trees, by their minute roots. To none of these can that motion be necessary to which plants are naturally exposed, and which, as has been already seen (p. 161.), is of so much importance to the healthy maintenance of their functions. Hence it is, that among fruit trees the Vine never suffers from being trained: indeed its anatomical structure is specially suited to such a mode of existence ; while all erect trees, of whatever kind, whose branches nature intended to be rocked by the storm, and perpetually waved by the currents of air to which they are exposed, in all cases suffer more or less.

One of the commonest and worst diseases induced by training is a gradual impermeability of tissue to the free passage of sap, which appears to stagnate, so that in time the branches become debilitated and juiceless: the obstruction to the flow of the sap tends to produce coarse shoots from various parts of the branches, and especially from the roots. The cause of this seems to be the too rapid deposit of the sedimentary matter of lignification *, and to be induced by want of motion and excessive exposure of the leaves and branches to the sun. The effect of the latter is to inspissate all the juices, and to promote their formation; while the former increases the evil by not keeping the fluids in rapid circulation : just as we know that a slow stream, from a muddy source, deposits its impurities much more copiously than a rapid stream. As this evil arises out of the operation of training, and seems to be inseparable from it, there will be no expectation of a remedy being discovered.

The increase of the saccharine quality of fruit is by no means an advantage in all cases; it improves the peach, the nectarine, the pear, and the plum, in which sweetness is the great object : but it deteriorates the apple and the apricot, which are chiefly valued for their peculiar mixture of acidity and sweetness.

The protection received in the spring by trees trained upon walls exposed to the sun, while it

* See Introduction to Botany, ed. 3. p. 3.

advances the period of flowering, at the same time
causes it to take place at a season when they are
not sufficiently secure from spring frosts; and
hence the necessity of protecting such plants artificially by coping, screens, bushes, curtains, and
other contrivances. It is on this account that the
utility of flued walls is so much diminished, and
that they are found, in practice, more valuable for
ripening wood in autumn, than for guarding blossoms in the spring.

CHAP. XV.

OF POTTING.

WHEN a plant is placed to grow in a small earthen
vessel like a garden pot, its condition is exceedingly different from that to which it would be naturally exposed. The roots, instead of having the
power of spreading constantly outwards, and away
from their original starting point, are constrained
to grow back upon themselves; the supply of
food is comparatively uncertain; and they are
usually exposed to fluctuations of temperature and
moisture unknown in a natural condition. For
these reasons, potted plants are seldom in such
health as those growing freely in the ground; but,

as the operation of potting is one of indispensable necessity, it is for the scientific gardener, firstly, to guard against the injuries sustainable by plants to which the operation must be applied; and, secondly, to avoid, as far as may be possible, exposing them to such an artificial state of existence. That the latter may be done more frequently than is supposed will be sufficiently obvious, when we have considered what the purposes really are that the gardener needs to gain by potting.

The first and greatest end attained by potting is, the power of moving plants about from place to place without injury; greenhouse plants from the open air to the house, and *vice versâ;* hardy species, difficult to transplant, to their final stations in the open ground without disturbing their roots; annuals raised in heat to the open borders; and so on: and, when this power of moving plants is wanted, pots afford the only means of doing so. It also cramps the roots, diminishes the tendency to form leaves, and increases the disposition to flower. Another object is, to effect a secure and constant drainage from roots of water; a third is, to expose the roots to the most favourable amount of bottom heat, which cannot be readily accomplished when plants of large size are made to grow in the ground even of a hothouse; and, finally, it is a convenient process for the nourishment of delicate seedlings. Unless some one of these ends is to be answered, and cannot be effected in a more natural manner, potting is better dispensed with.

That it may be advantageously dispensed with, in many cases, is evident from several facts more or less well known. The nurserymen prefer " pricking out " their delicate seedlings into pans, or movable borders, instead of pots; and they always thrive the better. In conservatories, the necessity of shifting plants from place to place may be often avoided; while, under judicious management, those which are planted in the open soil have greatly the advantage of others, both in healthiness and easiness of management; and there is no doubt that Pine-apples will succeed better unpotted, if planted freely in soil exposed to a *proper amount of bottom heat.* This was first asserted by Mr. Martin Call, one of the Emperor's gardeners at St. Petersburg (*Hort. Trans.*, iv. 471.), and has been since practised very successfully by others. In the year 1830, a pineapple, obtained by this treatment, weighing 9 lb. 4 oz., was sent to the King of England by Mr. Edwards, of Rheola; and the success of other growers, in the same manner, has been remarkable. (See *Hort. Trans.*, n. s., i. 388.)

The exhaustion of soil by a plant is one of the most obvious inconveniences of potting. The organisable matter in a soluble state, contained in a garden pot, must necessarily be soon consumed by the numerous roots crowded into a narrow compass, and continually feeding upon it. The effects of this are seen in the smallness of leaves, the weakness of branches, the fewness and imper-fect condition of flowers, &c.; and the gardener remedies them by applying liquid manure, by fre-quent shifting, or by placing his plants in *pan-feeders*, shallow earthen vessels containing manure, to which the roots have access through the holes in the bottom of a pot. It is, however, to shifting, more particularly, that recourse is had for reno-vating the soil; and this, if skilfully performed, without giving a sudden and violent shock to the plant, is probably the best means; because the roots are thus allowed more liberty of distribution, and the earth is kept more open (more permeable) than when consolidated by repeated applications of liquid manure. There is, however, a difficulty in shifting plants without injury to their roots, in the midst of full vegetation; and at such times the application of liquid manure is preferable, when the soil requires renovation.

It is not, however, by mere exhaustion that potted plants render the soil unfit for their sup-port. Every one knows that the soil of a farm will not bear, year after year, the same kind of crop, but that one kind of produce is cultivated on a piece of ground one year, and is succeeded by some other kind; which practice, in part, constitutes the important system of rotation of crops. Not, however, to refer to matters extra-horticultural, it is notorious that an apple orchard will not immediately succeed upon the site of an old orchard of the same kind of fruit, and that no amount of manuring will enable it to succeed; a wall border, in which fruit trees have been long grown, becomes at last insensible to manure, and requires to be renewed; and, not to dwell upon an undisputed fact, Dahlias do not " like " the soil in which Dahlias were grown the previous year. This class of phenomena cannot be explained upon the principle of soil being exhausted, because that exhaustion is made good and yet to no purpose, unless we assume that land contains something mineral which each species prefers to feed on, and which is not contained in manure. But the slender power of selection possessed by the roots of plants (35.) would be unfavourable to this supposition, even if it were open to no other objections. It has of late years been thought that the excretory functions of the root (39.) would explain the deterioration of soil, and that the reason why plants cannot grow year after year in the same soil, if it and their roots are disturbed, is, that, under such circumstances, they are perpetually brought into contact with the matter of which nature had previously relieved them; this matter being assumed to be unsuitable to themselves, although harmless to different species. The

subject has been hitherto so little investigated that
it is not safe, perhaps, to take it as the basis of a
theory; but it certainly appears to offer a more probable explanation of the deterioration of soil than
any other yet proposed. There are those, indeed,
who seem willing to deny altogether that soil is deteriorated; and cases are adduced of Peach trees
not repotted for twenty years, which did not die;
of Strawberry beds not renewed for a long series of
years, which still bore fruit: but I do not know
that any one ever asserted that trees would perish
if replanted in their own deteriorated soil; it has
only been said that they would become unhealthy
and unproductive, and I think few gardeners will
deny that. Neither has it been pretended that the
root-secretions of every plant are deleterious at all.
It is quite conceivable that one plant may secrete
a deleterious matter that is very slowly decomposable, but which may, nevertheless, be soluble
enough to enter into the food of other roots; and
in such a case an injurious effect may be produced:
while, in another case, the secreted matter may be
rapidly decomposable, when it will enter into new
combinations, and lose whatever deleterious property it originally possessed, if any. At all events,
be the theory what it may, it is an undoubted fact
that soil is deteriorated by a plant which has grown
in it for a long time; and that, to be maintained in
a healthy condition, that soil must be changed.

This explains why potted plants, carefully attended to and often shifted, are so much more healthy than those treated otherwise.   It is not, however, merely for the purpose of removing deteriorated earth or adding manure, that shifting is important; all potted plants have, in time, their ball of earth, by the continual passage of water through it, reduced to a state of hardness and solidity unfavourable to the retention of moisture or the growth of roots; and this is of course cured, if the operation of shifting is judiciously performed.   I must, however confess, I *have* seen gardeners contented with lifting a plant, with a hard old matted ball, out of one pot into another of a little larger size, shaking some particles of fresh earth in between the ball and the side of the pot, and pressing the whole down with as much force as the thumbs can give.

It is found that the roots of potted plants invariably direct themselves towards the sides of the pot, as must indeed necessarily happen in consequence of their disposition to grow horizontally. Having reached the sides, they do not turn back, but follow the earthenware surface, till at last they form an entangled stratum enclosing a ball of earth ; then, if not relieved by repotting, they rise upwards towards the surface, or they attempt to force themselves back to the centre.   The greater part, however, are always found in contact with the porous earthen side of the vessel; and especially

all the most powerfully absorbent, that is youngest,
parts. They are, therefore, in contact with a body
subject to great variations of temperature and moisture, in consequence of exposure to the sun, or to
a dry air in motion, unless in those rare cases where
the air is kept by artificial means shaded, and uniformly damp. By these means, in a dry summer
day, when the leaves are perspiring freely, and
therefore requiring an abundance of water from
the roots, the latter are placed in contact with a
substance whose moisture is continually diminishing; or in a greenhouse, where the pots are
syringed, the heat of the earth in contact with the
roots is lowered by a copious evaporation from the
sides of the pot, just when, in nature, the bottom
heat should be the greatest. The evil consequences
of this are well known to gardeners, who however
seldom take any sufficient precautions to prevent
it. Greenhouse plants exposed to the open air in
summer always suffer severely from the irregular
condition of the sides of the pots; whence the
common practice of plunging them in the earth,
for the purpose of bringing them into the condition
of plants growing in the open ground.

This is, however, attended with some disadvantage; for the plants root, through the bottom of the
pots or over the edges, among the earth in which
they are plunged; and, when taken up in the
autumn for removal into the greenhouse, they

must have all such roots cut off again; for there
are no means of bringing them within the limits of
a pot.   For these and similar reasons, no good
gardener will expose his greenhouse plants to the
open air in summer, *if he can help it ;* unless they
are duplicates, or unless there is some object to be
attained very different from the strange notion that
they are hardened by this process.   The effect that
is really produced upon them is, to give them a
sort of artificial winter in summer, that is, to expose them to a period of comparative rest from
growth, which, in many cases, is useful.

The best method of counteracting the injurious
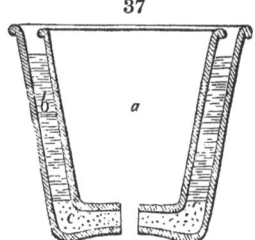 effects of exposure to the air is
by employing double pots (*fig.*
37.), as recommended in the
Gardener's Magazine, ix. 576.,
and by Captain Mangles, in his
Floral Calendar, p. 44.; the
space (*b*) between the two pots
being filled up with moss, or any other substance
retentive of moisture.

Of course the inconveniences now alluded to
are principally sustained by plants in small pots :
when the quantity of earth is considerable, as in
tubs or the largest kinds of pots, the loss of water
through the sides is of little moment; and the variation of temperature is more than counteracted by
the large surface exposed to the direct influence of

the solar rays. In these cases, the perfect drainage of superfluous moisture is often of the greatest service. Mr. Knight, indeed, assures us that "plants of every species are more or less affected, but not all injuriously, by having the sides of their pots exposed fully to the air. The taste and flavour of the peach and nectarine, and still more of the strawberry, are greatly improved; and the Fig-tree, in the stove, is made to afford a longer succession of produce, owing to the succession of young shoots, which are caused to spring from its larger branches and stems ; and, in all cases *when trees can be made to retain their health* in exposed pots, the period of the maturity of their fruit is very considerably accelerated." (*Hort. Trans.,* vii. 258.)

It seems to be nothing but the complete drainage to which they are then exposed, that makes the Orange and all its tribe, naturally inhabitants of the hill-sides of the temperate parts of Asia, thrive best when the roots come in contact with the sides of the pots, &c., in which they grow. In all cases, the drainage should be most carefully secured, by placing an abundance of broken tiles, potsherds, &c., in the bottom of a pot, so as to prevent the stagnation of water (page 123.) about the roots.

Mr. Macnab, in his excellent practical treatise upon the cultivation of Cape Heaths, points out

very forcibly the value of good draining to that
class of plants. There is scarcely any danger, he
says, of giving too much draining; and, in order
to effect this essential object still more perfectly,
he, in shifting his Heaths, constantly keeps the
centre elevated above the general level of the
earth in the pot or tub, so that at last each plant
stands on the summit of a small hillock.

In order to counteract the risk of excessive
drainage, without in reality diminishing it, great
advantage is derived from the introduction into
the earth of fragments of some absorbent stone.
Mr. Macnab uses " coarse soft free-stone broken
into pieces from one inch to four or five inches in
diameter ;" because in summer these stones retain
moisture longer than the earth, and in winter allow
a free circulation of any superabundant moisture.

If woody plants are allowed to remain growing
in the same pot for many years, as is sometimes
the case, one of two things must happen: either
the roots, matted into a hard ball, become so tortuous and hard as to be unfit for the free passage
of sap through them; or they acquire a spiral direction.   In either case, if such plants are turned
out of their pots in a conservatory, or in the open
ground, with a view to their future growth in a
state of liberty, new roots will be made with difficulty, and it will be a long time before the effects
of growth in the free soil will be apparent.   Where

the spiral or corkscrew direction has been once taken
by the roots, they are very apt to retain it during
the remainder of their lives ; and if, when they
have become large trees, they are exposed to a
gale of wind, they readily blow out of the ground,
as was continually happening with the Pinaster
some years ago, when the nurserymen kept that
kind of Fir for sale in pots. In all such cases as
these, the roots should be carefully disentangled
and straightened at the time when transplantation
takes place.

If, however, a potted plant is managed in the
most perfect manner, no such entanglement or
coiling up will take place. To be managed perfectly, a plant, when young, should be placed in as
small a pot as it will grow in, and it should be
gradually and successively transferred to larger
pots as it advances in size. If this is done, the
warmth to which the pot is exposed will be more
immediately felt by the roots ; the latter, as they
grow, will ramify regularly all through the mass
of earth, which, moreover, will be thoroughly
drained : but, if, on the other hand, a very small
young plant is placed at once in a large pot, and
left to grow there, the drainage will be less perfect,
the large mass of earth will be less sensible of the
heat to which it is exposed, the roots will from
the first take a horizontal direction towards the
outside of the pot, and, once there, will follow its

surface as has been already stated, exhausting the
small quantity of earth with which they are then in
contact, and profiting little or nothing by the main
body of soil in the interior of the pot. As the
proper manner of managing potted plants is of the
first consequence, I transcribe the following mode
of treating the Balsam, from a very sensible paper
by the Rev. William Williamson.

" As soon as they have got four leaves, I transplant them singly into the smallest pots I can procure, and in such a manner that the stem of the
plant may be covered somewhat more than it was
at first, and then all are to be again placed in the
frame. In a short time, if there be a sufficiency
of heat, that part of the stem which is covered
with the mould puts forth fibres, by which nourishment is conveyed more immediately to the principal stem of the plant. As soon as the plants are
a little advanced in growth, they are again removed (if possible without disturbing the earth)
into somewhat larger pots, still planting them
rather deeper than before. The same process is
repeated five or six times, till, at last, they are
removed into their final pots. I have found it
best to give them their last removal after they have
opened their first blossoms, as it gives additional
brilliancy and size to the flowers. By following
this method the plant acquires extraordinary vigour, throwing out its branches from the surface

of the mould, exhibiting flowers nearly as large as a full-blown rose, and a stem measuring two, and sometimes three, inches in circumference." (*Hort. Trans.*, iii. 128.)

The plan of continually sinking the stem with every succeeding potting is useful to the Balsam, because it puts forth roots in abundance from its stem ; and to all plants having the same property, the same practice is desirable : but not to others, which, if their stems do not root as fast as they are buried, will suffer injury by the sinking.

It is by paying constant attention to the shifting of the growing plant, by the employment of a very rich stimulating soil, and by a thorough knowledge of the kind of atmosphere which suits them best, that have been obtained those magnificent Pelargoniums, Cockscombs, Balsams, and similar flowers, which have so often and so justly excited the admiration of even the most experienced gardeners.

CHAP. XVI.

OF TRANSPLANTING.

As soon as man attempted to beautify his residence with trees planted round it, he would naturally obtain them from the forest ; and he then would find that, of many that he removed, all or some at least would die : if however he persevered, he would at last discover that while constant failure attended his efforts at one time, comparative success would crown them at another ; and he would thus be led to investigate, according to his skill, the causes of success and failure. Out of this would grow in time the art of transplanting, among the most important business of the gardener.

I fear, however, it is too generally practised as an empirical art, without sufficient attention being paid to the principles on which its success or failure depend ; at least, one hardly knows how to draw any other conclusion from the opposite opinions held by planters, the dogmatical manner in which they are too often expressed, and the obscure and unintelligible phraseology of what are called explanations of the practice by amateurs, to whom it is not necessary to allude more particularly. If there is any one part of the art of Horticulture in which *post hoc* has been mistaken for *propter hoc* more commonly than another, it is surely in what concerns transplantation.* And

* It is scarcely necessary to say that these remarks *do not*, in any way, apply to Mr. Macnab's *Hints on the Planting and general Treatment of Hardy Evergreens in the Climate of Scotland ;* an excellent treatise, which it it impossible to recommend too strongly to the attention of the planter.

yet the rationale is simple enough; if we do not labour to render it confused by imaginary refinements.

When a plant is taken out of the ground for transplanting, its roots are necessarily more or less injured in the process, and consequently it is less able to support the stem than it was before the mutilation took place; its loss of this power will also be in proportion to the extent of the mutilation, which may be carried so far as to amount to destruction.

But the importance of their roots to plants is not alike at all seasons; in the summer, when there is the greatest demand upon them in consequence of the perspiration of the foliage (70., &c.), they are most essential; in winter, when the leaves have fallen, they are comparatively unimportant, as is evident from a very common case. Let a limb of a tree be felled in full leaf in June; its foliage will presently wither, the bark will shrivel and dry up, and the whole will speedily perish; but, if a similar limb is lopped in November, when its foliage has naturally fallen off, it will exhibit no sign of death during winter, nor till the return of spring, when it may make a dying effort to recover; but the means it takes to do so, namely, the emission of leaves, only accelerates its end.

These two propositions really include all the most essential parts of the theory of transplantation, as will presently be seen : it is necessary, however, that they should be applied in some detail; for which purpose it will be convenient to consider, first, the *season*, and, secondly, the *manner*, in which transplanting can be best effected.

It is the powerful perspiratory action of the leaves of deciduous trees which renders transplanting them in a growing state so difficult, that for practical purposes it may be called impossible; for the operation is necessarily* attended by a mutilation of the roots which feed the leaves.   At no period, then, can the operation be performed if such plants are growing.   Even if the buds are only pushing, the process should be avoided, because immediately after that period the demand upon the roots is greatest ; for although in consequence of the smallness of the surface of the young leaves the action of perspiration may seem to be feeble, yet the thinness of the newly formed tissue will not enable it to resist the drying action of the atmosphere unless there is a most abundant afflux of sap from the roots.   In England, too, the months when buds begin to burst forth are objectionable, not only on account of their dryness (see the tables at page 136.), but of their coldness, which prevents

* Transplanting from garden pots, in which the roots are preserved artificially from injury, may be performed equally well at any time if care is taken, and is, of course, not included in this statement.

the free circulation of sap ; and their evil effects are felt not only by the roots through the foliage, but directly, as will be shown hereafter. The season, then, which ought to be chosen is the period that intervenes between the fall of the leaf in autumn and the earliest part of spring, before the sap begins to move and the dry cold winds of that season to prevail. I entirely agree with Mr. Macnab, that the earliest time at which planting can be effected is, upon the whole, the best; a conclusion to which he has come from his extensive practice, in which my own observation of a great deal of planting for the last twenty-five years coincides, and which is, in all respects, conformable to theory. As soon as a plant has shed its leaves it is as much at rest for the season as it will be at any subsequent period, unless it is frozen ; its torpor, indeed, is greater at that time, because its excitability is completely exhausted by the season of growth, and it has had no time to recover it. If, at that time, a root is wounded, a process of granulation or cicatrisation will commence, just as it does in cuttings (page 210.) ; and from that granulation, which is a mere developement of the horizontal cellular system (45.), roots will eventually proceed. Now, it is obvious that since roots *must* be wounded in the process of transplantation, the sooner the wound is made the better, because it has the longer time in which to heal : and therefore the earlier in the autumn transplanting is effected, the less injury will be sustained by the plant submitted to the process; in the technical language of the gardener, "it has the more time to establish itself."

Autumn and mid-winter are, moreover, the best seasons, because of their great dampness. It will be seen by reference to Mr. Thompson's tables (page 136.), that the air is very generally in a state of saturation in the months of October, November, December, January, and February, and that it is seldom in that condition at any other season. Now, although the perspiration of plants is greatly diminished by the removal of the leaves, it is not destroyed, for they also perspire through their young bark; and therefore a saturated atmosphere, which prevents much of the perspiratory action which remains from being exercised, is a condition, even when plants are leafless, much too beneficial to be overlooked. Nor is the action upon the perspiratory power of the stem the only mode in which a saturated atmosphere is important at the time of transplantation; it exercises a directly favourable influence on the roots themselves. Roots at their spongioles, or most absorbent points, are extremely delicate parts, unprotected by a fully organised epidermis (22.), destined to exist in a moist medium, and capable of being easily killed by exposure to dryness as well as by actual violence. The accidents to which the roots of transplanted trees are liable, from the very nature of the operation, are of such a kind that it is impossible to prevent their being exposed to the air, sometimes for considerable periods of time; it is therefore obviously a point of the first importance, that the air should be as nearly of the humidity of the soil from which the roots have been extracted as can be secured. How unfavourable, in this point of view, the months of March, April, and May are for planting, is apparent from Mr. Thompson's tables above referred to; how little the matter is attended to by nurserymen, gardeners, and labourers, all great planters know to their cost. Mr. Macnab, who thoroughly understands all this, prefers a moist rainy day; although, as he says, he has " at times been as wet in planting evergreens, as when exposed for hours on the windy side of Ben Nevis in a wet day, without great coat and with a broken umbrella." It may be very true that good plantations have been made in March and April; it may be equally true that no such care as I have described is necessary for all plants; but no wise man would, on that account, neglect the precautions which the nature of plants shows to be necessary to insure success with all things. Very wet and late springs may prevent the loss of any considerable proportion of the trees planted in March and April, especially if succeeded by a dull, warm, wet summer; and a Willow may be planted with success at midsummer: but we cannot tell beforehand what sort of spring is coming, and all plants have not the tenacity of life possessed by a Willow.

If the months of November and December are the most favourable for transplanting deciduous trees, and March and April the worst, how much more important must be those periods to evergreens. An evergreen differs from a deciduous plant in this material circumstance, that it has no season of rest; its leaves remain alive and active during the winter, and consequently it is in a state of perpetual growth. I do not mean that it is always lengthening itself in the form of new branches, for this happens periodically only in evergreens, and is usually confined to the spring; but that its circulation, perspiration, assimilation, and production of roots are incessant. Such being the case, an evergreen, when transplanted, is liable to the same risks as deciduous plants in full leaf, with one essential difference. The leaves of evergreens are provided with a thick hard epidermis (61.), which is tender and readily permeable to aqueous exhalations only when quite young, and which becomes very firm and tough by the arrival of winter, whence the rigidity always observable in the foliage of evergreen trees and shrubs. Such a coating as this is capable, in

a much less degree than one of a thinner texture,
such as we find upon deciduous plants, of parting
with aqueous vapour; and moreover its stomates
(61.) are few, small, comparatively inactive, and
chiefly confined to the under side, where they are
less exposed to dryness than if they were on the
upper side also. But although evergreens, from
their structure, are not liable to be affected by the
same external circumstances as deciduous plants,
in the same degree; and although, therefore, transplanting an evergreen in leaf is not the same
thing as transplanting a deciduous tree in the
same condition, yet it must be obvious that the
great extent of perspiring surface upon the one,
however low its action, constitutes much difficulty,
superadded to whatever difficulty there may be in
the other case. Hence we are irresistibly driven
to the conclusion, that whatever care is required in
the selection of a suitable season, damp, and not
too cold, for a deciduous tree, is still more essential
for an evergreen. It is, therefore, most extraordinary, that it should have ever been the practice to defer the planting evergreens till late in the
spring, upon the supposition that it is the very
best season for them, and that midsummer even is
a proper period; as if cold winds, accompanied by
from 20° to 30° of dryness in the air, which is not
more than 500 or 357 of moisture, with a bright
sun beating on the roots which are exposed, and

exciting the action of the perspiring surface to the utmost extent of its power, were external conditions with which the gardener has no concern : and yet, as Mr. Macnab justly observes, half a day's sun in spring and autumn will do more harm immediately after planting, than a whole week's sun from morning to night in the middle of winter.

The Holly, says a writer in the *Horticultural Transactions*, does not succeed well, if transplanted at any other season of the year than the end of April or beginning of May ; at this time the buds are just breaking open into leaf, and I have rarely failed of success in transplanting small, or even very large old, trees. (ii. 357.) Although such statements cannot be too strongly contradicted as guides to practice, yet it is not difficult to explain their origin. As evergreens are never deprived of their leaves, so they are never incapable of forming roots ; on the contrary, they produce them abundantly all winter long, and rapidly at any other period of the year which is favourable to their growth : so that they are capable of making good an injury to their roots much more speedily than deciduous plants ; especially as in the majority of cases the roots are numerous and fibrous, and not so liable to extensive mutilation when transplanted. Now, if an evergreen is planted in the month of May, and the weather *happens* to be cloudy, mild, and damp, as the plant is just then commencing the renewal of its growth, and is forming fresh roots abundantly, if such a state of weather lasts for a week or two, there is no doubt that the plant will succeed very well; and so it will if removed at midsummer. In the year 1822, in the month of August, there were planted in the garden of the Horticultural Society of London above 6000 Hollies from two to three feet high, for the purpose of forming fences : few plants in all that number ever exhibited any traces of having been removed, and I do not believe that a hundred died. The weather was dry ; but the plants were deluged with water when placed in their holes, and they had been obtained from the Regent's Park, where they grew in the stiff plastic clay of that side of London ; the consequence of which was, that, when taken out of the ground, so much earth adhered to them, that they were almost in the state of plants removed from pots. Now, is this a case to justify planting Hollies in the month of August ? Surely not ; it only shows that it may be done under a combination of very propitious circumstances. There may be local conditions of a permanent nature, owing to the peculiarity of climate, in which those advantages may be calculated upon ; but they do not justify the gardener in taking a season of great risk, instead of a season of perfect certainty. I have seen tens of thousands of Hollies planted late in the spring in the county of Norfolk, and in the quarters, too, of nurseries, where, from the plants shading each other, they are far more likely to succeed than if exposed singly; and although it sometimes happened that a good many lived, it is not too much to say that three fifths at least would die; and it is perfectly well known that if planted in the beginning of November no such loss is sustained. In short, I am certain that if experience is looked to only, it will give the same answer as theory to the question of what season is the best for planting evergreens, namely, that which is best for other trees; and such cases to the contrary as may appear to exist will always be found exceptions to the rule, in consequence of some peculiar circumstances attending them; not unfrequently, I believe, from the operation having been performed upon a very small number of plants, to the removal of which a degree of care was given wholly incompatible with general and extensive practice.

Mr. Macnab rightly adverts to the importance of choosing a suitable day, as well as season, for the operation; and it must be evident from what has now been stated, that this is very necessary : as, however, the theory of this is the same as that of the season, it will be sufficient to quote this excellent practical gardener's rules. In winter, you may plant with perfect safety in a dull calm day, whereas in spring or autumn a moist rainy day is preferable to any other; but where a person has not the choice of such weather, then the work should be performed in the evening, when the sun gets low, especially in spring or autumn planting. (p. 22.)

Next in importance to the selection of a fitting season, is the preservation of the roots of transplanted trees; the former is of little consequence, if the latter is not most carefully attended to. We know, indeed, that some plants will live with the rudest treatment, and bear the most severe mutilation without much suffering; but those are special instances of extreme tenacity of life, and do not affect general principles. The value of great attention to the roots, in the operation of shifting, has already been pointed out (p. 283.), and transplanting is only shifting in another manner. It would be the duty of the gardener to save every minute fibre of the roots, if it were practicable; but, as that is not the case, his care must be confined to lifting his trees with the least possible destruction of those important organs; remembering always that it is not by the coarse old woody roots that the absorption of food is carried on, but by the youngest parts, and especially the spongioles (23, 24.). The mechanical means by which this is best effected do not belong to the present subject; I may however remark, without quitting the limits of theory, that, as the greater part of the young fibres is produced at the circumference of the circle formed by the root, the earth should be first removed at some distance from the stem, so as to insure, as far as possible, their being taken up entire; if this is not done, but the spade is struck into the earth near the stem, or if the rude nursery practice, justly enough called drawing, is employed, a large part of the most valuable roots must necessarily be cut off or destroyed by tearing. The greatest difficulty, beyond that of mechanical removal, in transplanting trees of considerable size, is this preservation of roots; and, if it were possible to carry without injury such heavy masses as old forest trees, there is no physical obstacle to transplanting them, if the extrication of the fibrous part of the roots be secured, which is not impracticable. As, however, the latter is a troublesome and very difficult operation, even when trees are only ten or twelve feet high, it has been, from time out of mind, the custom of skilful planters to prepare such trees for removal by cutting back their main roots one year before they are to be transplanted; if this very simple operation is properly performed, all the principal limbs, so amputated, will emit young fibres in abundance from their extremities, and the gardener, from knowing where to find those roots, can easily take them up without material injury. In order to effect the same end, but in another way, the following expedient has been occasionally employed for large trees. A deep trench has been opened, in midwinter, round a stem, at such a distance as to be clear of the principal fibres; the tree has then been carefully undermined, till, at last, the earth belonging to it has formed a huge ball; upon the approach of frost, water has been freely poured over the ball so that its whole surface may be converted into an icy mass; in that state it has been raised by powerful tackle, and conveyed without disturbance to its intended site. This operation, which is the best possible for hardy trees of great size, but expensive, and therefore only capable of application in a limited degree, owes its success entirely to the young and tender fibres being placed in such a position that they cannot be injured by the act of transport.

Under all ordinary circumstances, the roots must necessarily be injured more or less by removal; in that case, all the larger wounds should be cut to a clean smooth face; not in long ragged slivers, as is often the case, and which is only substituting one kind of mutilation for another, but at an angle of about 45°, or less. If the ends of small roots are bruised, they generally die back a little way, and then emit fresh spongioles; but the larger roots, when bruised, lose the vitality of their broken extremity, their ragged tissue remains open to the uncontrolled introduction of water, decays in consequence of being in contact with an excess of this fluid, and often becomes the seat of disease which spreads to parts that would otherwise be healthy. When, however, the wound is made clean by a skilful pruner the vessels all contract, and prevent the introduction of an excess of water into the interior ; the wound heals by granulations formed by the living tissue, and the readiness with which this takes places is in proportion to the smallness of the wound. It may be sometimes advantageous to remove large parts of the coarser roots of a tree, even if they are not accidentally wounded when taken up, the object being to compel the plant to throw out, in room of those comparatively inactive subterranean limbs, a supply of young active fibres. This is a common practice in the nurseries in transplanting young Oaks and other taprooted trees, and is one of the means employed by the Lancashire growers of Gooseberries, in order to increase the vigour of their bushes; in the last case, however, the operation is not confined to the time when transplantation takes place, but is practised annually upon digging the Gooseberry borders. The reason why cutting off portions of the principal roots causes a production of fibres appears to be this : the roots are produced by organisable matter sent downwards from the stem (31.) ; that matter, if uninterrupted, will flow along the main branches of the roots, until it reaches the extremities, adding largely to the wood and horizontal growth of the root, but increasing, in a very slight degree, the absorbent powers : but if a large limb of the roots is amputated, the powers of the stem remaining the same, all that descending organisable matter which would have been expended in adding to the thickness of the amputated part, is arrested at the line of amputation ; and, unable to pass further on, rapidly produces granulations to heal the wound, and immediately afterwards young spongioles, which soon establish themselves in the surrounding soil, and become the points of new active fibres.

The question of pruning the branches of transplanted trees has been already sufficiently adverted to (see p. 260.).

By many excellent planters, the advantage of deluging the roots with water, when newly planted, is much insisted on ; and in the case of large plants, particularly evergreens, it is, undoubtedly, an essential process, partly because it causes the flagging and injured roots to be immediately surrounded by an abundant supply of liquid food, which, if the operation be skilfully performed (see Macnab's *Treatise*, p. 24. and 25.), will not subsequently fail them ; and partly because it is the only means we possess of embedding with certainty all the fibres in soil. When the earth is reduced to the state of puddle, it will settle round the finest

roots, and place them as nearly as possible in the
same condition, with regard to the soil, that they
were in before the plants were removed.   But the
operation of puddling is unnecessary to small plants,
if removed at a proper season of the year, especially to deciduous trees of all kinds; and it may
be very injurious.   This was long ago stated by
Mr. Knight (*Hort. Trans.*, iii. 159.), who found
by experience that when trees are very much out
of health, in consequence of having become dry,
excess of moisture to the roots is often fatal.   This
appears to arise from the languid powers of the
plant being insufficient to enable it to decompose
and assimilate the water rapidly introduced into
its system through the wounds in its root, and by
the hygrometrical force of that part; under such
circumstances, water will dissolve the mucilaginous
and other matters intended for the support of the
nascent buds, which matters then putrefy, lose
their nutritive quality, and rapidly destroy the
tissue.   The substitute for root-watering contrived
by Mr. Knight in such cases was, to keep the
plants in a situation shaded from the morning sun,
and to moisten their bark frequently; by these
means, water is presented to them very slowly
through the young cortical integument (48.),
which, partaking of the nature of a leaf (63.),
slowly absorbs it, probably decomposes it, and
transmits it laterally (57.) through the liber into

the alburnum, where it finds itself in the ordinary channel for the ascending sap, and thus enters the system of circulation. In this way Mr. Knight originally preserved American Apple trees, which reached him in the middle of April, in so bad a state that they seemed " perfectly lifeless and dry, and much better fitted for fire-wood than for planting."

CHAP. XVII.

OF THE PRESERVATION OF RACES BY SEED.

THE manner of preserving the domesticated races of plants by the ordinary means of propagation, such as cuttings, layers, grafts, and so on, has already been explained ; there are, however, some other topics connected with this important subject which require to be touched upon.

Propagation by division is inapplicable to annuals or biennials, or at least can be practised upon only a very limited scale, and for such plants the gardener has to trust to seeds alone. But it is an axiom in vegetable physiology that seeds reproduce the species only, while buds (that is, propagation by division) will multiply the variety ; and this is undoubtedly true as a general rule. But the skill and care of the gardener often enable him to perpetuate by seed the many races of cultivated annuals, varieties of the same species, improved and altered by centuries of domestication, with as much certainty as if he were operating with cuttings. In a well managed farm we see the various breeds of Turnips and Corn preserving each its own peculiar character unchanged year after year, and yet they must all be propagated by seed alone; and in gardens the varieties are innumerable of Peas, Lettuces, Cabbages, Radishes, &c., whose purity is maintained by the same means. The manner in which this is effected is of the first importance to be understood.

Although it is the general nature of a seed to perpetuate the species only to which it belongs, and it cannot therefore be relied upon, in ordinary cases, to renew a particular variety of the species, yet there is always a visible tendency in it to produce a seedling more like its parent than any other form of the species. Suppose, for example, the seed of a Ribston Pippin apple were sown; if untainted by intermixture with other varieties, it would produce an apple tree whose fruit would be large, sweet, and agreeable to eat, and not small, sour, and uneatable like the Wilding Apple or Crab. The object of the gardener is to fix this tendency, and he does it by means not unlike those employed in the preservation of the races of domesticated animals, namely, by "breeding in and in," as the phrase is.   An example of this will be more instructive than a dissertation. The Radish has, when wild, a long pallid root; among many seedlings one was remarked with roots shorter and rounder, and more succulent than the remainder; this was a "sport" to which all plants are subject.   Had that Radish been left among its companions, and the seed saved from them all indifferently, the tendency would have disappeared for that time; but its companions were all eradicated, and the better one produced its seed in solitude.   The crop of young plants obtained from this Radish was, for the most part, composed of individuals of the wild form, but several preserved the same qualities as the parent, and some, perhaps one only, in a higher degree: in this one, then, the tendency was beginning to fix.   Again were all eradicated, except the last-mentioned individual, whose seeds were carefully preserved for the succeeding crop; and, by a constant repetition of this practice for many years, at last the habit to produce a round and succulent root became so fixed, that all the Radishes assumed the same appearance and quality, and there were none left to draft or "rogue." Every variety of annual crop, not still in its wild state, must have gone through this process of

fixing; and thus the varieties of earliness, late-
ness, and productiveness, colour, form, and flavour
observable in garden plants, have been secured for
our enjoyment.

But to fix a new habit in annual plants is not
the only care of the cultivator, whose patience
and skill would be ill employed if it could not be
preserved. If a plant has some tendency to vary
from its original condition, it has much more to
revert to its wild state; and there can be no doubt
that, if the arts of cultivation were abandoned for
only a very few years, all the annual varieties of
our gardens would disappear, and be replaced by a
few original wild forms.

For the means of preserving the races of plants
pure, the means vary according to the nature of
the variety. As far as concerns early and late
varieties, it often happens that, as in Peas, the
tendency in such plants to advance or retard their
season of ripening was originally connected with
the soil or climate in which they grew. A plant
which for years is cultivated in a warm dry soil,
where it ripens in forty days, will acquire habits
of great excitability; and, when sown in another
soil, will, for a season or so, retain its habit of
rapid maturity: and the reverse will happen to an
annual from a cold wet soil. But, as the latter will
gradually become excitable and precocious, if sown
for a succession of seasons in a dry warm soil, so

will the former lose those habits, and become late
and less excitable. Hence, the best seedsmen
always take care that their early varieties of annuals are procured from warmer and drier lands
than those on which they are to be sown; our
earliest Peas, for example, are obtained from France,
and the next in time of ripening from the hot dry
fields of Kent, the Suffolk coast, and similar situations. Thus, also, the Barley grown on sandy
soils, in the warmest parts of England, is always
found by the Scotch farmer, when introduced into
his country, to ripen on his cold hills earlier than
his crops of the same kind do, when he uses the
seeds of plants which have passed through several
successive generations in his colder climate; and
Mr. Knight found that the crops of Wheat on
some very high and cold ground, which he cultivated, ripened much earlier when he obtained his
seed-corn from a very warm district and gravelly
soil, which lies a few miles distant, than when he
employed the seed of his vicinity. It would seem
as if this were in some way connected with the
mere size of a seed, the smallest seeds of a given
variety producing plants capable of fructifying
quicker than those of a much larger size. We
have, at present, but little information upon this
subject; but there are some most curious experiments relative to it by Messrs. Edwards and Colin,
who found that, although Winter Wheat cannot, in

France, be made to shoot into ear, if sown in the spring, provided the largest grains of the variety are employed, yet that, if the smallest grains are picked out, some will ear like Spring Wheat (see *Annales des Sciences Naturales*, v. 1.). Out of 530 grains of Winter Wheat, sown on the 23d of April, and weighing 7 ounces 52 grains, not one pushed into ear, they tillered abundantly, but the tillers were excessively stunted, and concealed among the tufts of leaves ; in short, they formed nothing but turf : on the other hand, of 530 other grains, weighing 3 ounces 56 grains, and sown on the same day, 60 pushed in ear.

It would seem as if many of our most esteemed garden plants were the result of debility, and that the succulence, the sweetness, or the excessive size, which render them so well suited for food, were only marks of unhealthiness. At least, it is almost necessary to assume this to be the case, in order to account for the efficacy of one of the modes of maintaining races genuine. It is perfectly well known, that, if such an annual as a Turnip is transplanted shortly before it runs to seed, the characters of its variety will remain more strongly marked, and have far less tendency to vary, than if, all other circumstances remaining the same, the seed is saved without the process of transplantation having been observed. Now, the only effect of transplanting, at the season immediately preceding the formation of a flower-stalk, would seem to be that of checking the luxuriance of the individual operated on ; or, upon the above assumption, of increasing its debility of constitution. And the same explanation appears applicable to a strange custom mentioned by Mr. Ingledew as being practised in the Dekkan, to prevent the rapid deterioration, in that climate, of the Carrot, the Radish, and the Parsnep, the favourite table vegetables of the inhabitants. He states that the Indian gardeners, in the first place, prepare a compost of buffaloes' dung, swine's dung, and red maiden earth, mixed with water till they have the consistence of paste, and scented with a small quantity of assafœtida, the latter of which seems to be perfectly useless.

" The vegetables for this operation are drawn, when wanted, from the beds, when they have attained about one third of their natural growth, and those plants are chosen which are the most succulent and luxuriant; the tops are removed, leaving a few inches from their origin in the crown upwards; and a little of the inferior extremity, or taproot, is cut straight off likewise, allowing nearly the whole of the edible part to remain, from the bottom of which to within about an inch of the crown, are made two incisions across each other entirely through the body of the vegetable, dividing it into quarters nearly to the upper end.

They are then dipped into the compost until they are well covered by it, both externally and internally, and are immediately placed in beds, previously prepared for their reception, at the distance of fifteen or sixteen inches from each other, and so deep in the ground that the upper extremities only appear in sight. They are afterwards regularly watered; and when they take root, and fresh tops have made some advance in growth, they require but little attention. The tops speedily become large, and grow into strong and luxuriant stalks, the blossoms acquire a size larger than ordinary, and the seed they produce is likewise large and vigorous, and superabundant in quantity. Innumerable roots are thrown out from the incised edges of these plants; they consequently receive a greater abundance of nourishment, which occasions their luxuriant growth, causes them to yield not only a more than ordinary crop of seed, but also of a superior quality. (*Hort. Trans.*, v. 517.) The operation is performed at the beginning of the dry season.

Besides "roguing out" (i. e. eradicating) all individuals having the slightest appearance of degeneracy from among the plants intended for seed, care must be taken that the crop is so far from any other of a similar kind as to incur no risk of being spoiled by the intermixture of its pollen (88.). This substance is conveyed to considerable distances by wind and insects; and it is scarcely possible to be secure from its influence, if similar crops are cultivated within some miles of each other; whence we find certain villages, in different parts of Europe, celebrated for the purity of the seed of particular varieties; this usually happens in consequence of the villagers cultivating that variety and no other, as happens at Castelnaudary with Beet, at Altringham with the Carrot, and in Norfolk with different kinds of Turnip.

It is, however, to be observed, that the deterioration of seed by bastardising happens to a greater extent to single plants than to large masses of them; and it seldom happens that good seed can be saved in a garden, or near gardens, from a single individual. Solitary specimens of the Turnip, the Cauliflower, and such plants, have been frequently selected on account of their perfect characters, and been carefully planted in gardens for a stock of seed, but their produce has as frequently been of the worst description, bearing no resemblance to the parent. In such cases as these, it would seem as if bees and other insects were attracted from all quarters by the gay colours, or odour, of such isolated individuals, and, arriving from a hundred flowers which they had previously visited, bring with them so many sources of contamination.

When, however, the action of other flowers can

be prevented, as in the Melon and other unisexual plants, by "setting," the largest, healthiest, and most cultivated varieties will yield seed of the purest and finest quality. The tendency of Persian Melons to degenerate in this country was remarked soon after their introduction; and, for a long time, it was thought impossible to preserve them for many generations. Mr. Knight, in his numberless experiments upon this fruit, found that to be the case; for his fruit, at one time, became less in bulk and weight, and deteriorated in taste and flavour. But when he came to consider that " every large and excellent variety of the Melon must necessarily have been the production of high culture and abundant food; and that a continuance of the same measures which raised it to its highly improved state must be necessary to prevent its receding, in successive generations, from that excellence;" the cause of his Persian Melons deteriorating became apparent: and he found that by bringing the cultivation of the plants to a state of great perfection, he succeeded completely in rendering the original quality hereditary, as long as those precautions were observed. No man was more successful in the cultivation of the Melon than Mr. Knight; and it is in the memory of many persons, that the quality of his Sweet Melons of Ispahan has very rarely been equalled. The peculiar methods that he adopted appear to have been the complete and most careful preservation of the leaves from injury of whatever kind, the full exposure of their surface to light, and the augmentation of the ordinary warmth of a Melon bed by availing himself of the heat reflected from brick tiles with which his bed was paved. To such an extent was his care of the leaves carried, that he would not allow even the watering to be performed " over head," but he caused his gardener to pour water, from a vessel of proper construction, upon the brick tiles between the leaves, without touching them. (See various papers upon the Melon in the *Horticultural Transactions*, and especially that in vol. vii. p. 584.)

While, however, such are the general principles upon which the preservation of the peculiar qualities of the many races of cultivated annuals necessarily depends, it must be confessed that, according to report, there are circumstances upon which science can throw no light, and which, if true, must depend upon conditions as yet unsuspected to exist. Of this class is the following, respecting the Brussels Sprouts Cabbage, given upon the authority of M. Van Mons.

" Much has been said of the disposition of this plant to degenerate. In the soil of Brussels it remains true, and I have lately observed it to do the same at Louvain ; but at Malines, which is the same distance from Brussels as Louvain, and where

the greatest attention is paid to the growth of vegetables, it deviates from its proper character, after
the first sowing : yet it does not seem that any
particular soil or aspect is essential to the plant,
for it grows equally well and true at Brussels, in
the gardens of the town, where the soil is sandy
and mixed with a black moist loam, as in the fields,
where a compact white clay predominates.   The
progress of deterioration at Malines was most
rapid ; the plants raised from seed of the true sort,
which I had sent there, produced the sprouts in
little bunches or rosettes, in their true form; seeds
of those being saved, they gave plants in which
the sprouts did not form into little cabbages, but
were expanded ; nor did they shoot again at the
axils of the stem.   The plants raised from the
seeds of these last mentioned only produced lateral
shoots with weak pendent leaves, and tops similar
to the shoots, so that in three generations the
entire character of the original was lost.   From a
plant in the state last described, seed was saved at
my request, and sent back to me.   I had it sown
by itself, and carefully watched the plants in their
growth ; I was not long in discovering that they
retained the same character of degeneration they
had assumed at Malines, and preserved it throughout the whole course of their growth, yielding
pendulous leaves with long petioles, and having
no disposition to cabbage.   I suffered these plants

to run to seed at a great distance from my true Sprouts, which the extent of my garden allowed me easily to do. The second sowing brought them back a good deal to their true character; the plants yielded small cabbages regularly at each axil, but not generally full or compact, and they did not shoot a second time, as the true sort does. I again suffered these to run to seed, using the same precaution of keeping them by themselves. I sowed the seed, and this time the plants were found to have entirely recovered their original habits, their head, and rich produce." (*Hort. Trans.*, iii. 197.) I must confess, however, that, although the passage merits quotation, for the sake of exciting attention to the subject, it appears to me very doubtful whether the case has been fully, if correctly, stated.

CHAP. XVIII.

OF THE IMPROVEMENT OF RACES.

WHAT has been stated in the preceding chapter, concerning the preservation of the races of domesticated plants, is in some measure applicable to their improvement; because the very means employed to preserve those peculiarities of habit, which render them valuable, will, from time to time, be the cause of still more valuable qualities making their appearance. There are, however, other points of great importance on which the gardener has dependence.

A fixed improvement in the quality of the produce of a plant can only be obtained in one of two ways; either *directly*, by accidental variations in itself, or *indirectly*, by the process of muling.

Direct alterations in the quality of seedling plants often occur from no apparent cause, just as those accidental changes, called " sports," in the colour or form of the leaves, flowers, or fruit, of one single branch of a tree, occasionally break out, we know not why. Of these things, physiology can give no account; but it is known that, when such sports appear, they indicate a permanent constitutional change in the action of the limb thus affected, which changes may be sometimes perpetuated by seed, and always by propagation of the limb itself, when propagation is practicable. It is in this way that many of our fruits have probably, and several of the Chinese Chrysanthemums have certainly, been obtained. It was apparently thus that the Nectarine emanated from the Peach. It is possible that many new forms of shrubs might be procured by keeping these facts in view, and that climbers might be deprived of their climbing habits; for it is known that the handsome evergreen bush called the Tree Ivy, which grows erect, with scarcely the least tendency to climb, has been procured by propagating the fruit-bearing branches of trees of considerable age.

But we are by no means destitute of the power of procuring, with considerable certainty, improved varieties, by an application to practice of physiological principles. In the last chapter has been shown the importance of securing the production of seed by plants in the most healthy state possible, because a robust parent is likely to afford a progeny of similar habits to itself. In annuals, however, this is apparently restrained within narrower limits than in woody plants, from the great difficulty of fixing a new peculiarity in the former, and the facility with which it may be effected in the latter case, by means of buds, cuttings, grafts, and similar modes of propagation. The great object of the scientific gardener who desires to improve the varieties of plants upon principle will be, then, by artificial means, to bring the parent from which seed is to be saved as near as possible to that state at which he desires the seedling to arrive.

It is well known that the abstraction of fruit and flowers augments the vigour of the branches, or of the parts connected with them, and that the removal from the former of any part which takes up a portion of the food employed in the support of the flowers increases their efficiency. Thus those

varieties of the Potato, which will neither flower
nor fruit otherwise, may be made to do both by
stopping the developement of tubers ; and, on the
other hand, the size and weight of the tubers themselves are increased by preventing the formation of
flowers and fruit.   The course, then, to take, in
obtaining the largest possible tubers in a new
variety of the Potato, would be, in the first place,
to effect that end temporarily, but during several
successive seasons, by abstracting all the flowers
and fruit, and by such other means as may suggest themselves ; and then to obtain the most perfect seed possible by a destruction of the tubers
during the season when seed is finally to be saved.
Mr. Knight found, in raising new varieties of the
Peach, that, when one stone contained two seeds,
the plants these afforded were inferior to others.
The largest seeds, obtained from the finest fruit,
and from that which ripens most perfectly and
most early, should always be selected (*Hort. Trans.*,
i. 39.) ; and, in his incessant efforts to obtain new
varieties of fruit of other genera, he had reason to
conclude that the trees, from blossoms and seeds
of which it is proposed to propagate, should have
grown at least two years in mould of the best
quality ; that during that period they should not
be allowed to exhaust themselves by bearing any
considerable crop of fruit ; and that the wood of
the preceding year should be thoroughly ripened

(by artificial heat when necessary) at an early period in the autumn ; and, if early maturity in the fruit of the new seedling plant is required, that the fruit, within which the seed grows, should be made to acquire maturity within as short a period as is consistent with its attaining its full size and perfect flavour. Those qualities ought also to be sought in the parent fruits, which are desired in the off-spring ; and he found that the most perfect and vigorous progeny was obtained, of plants as of animals, when the male and female parent were not closely related to each other. (See the *Horticultural Transactions*, i. 165.)

There are no processes known to the cultivator so efficacious in producing new varieties as that adverted to in the last paragraph, that is to say, muling or cross breeding (88.) ; and it is to these operations, more than to anything else, that we owe the beauty and excellence of most of our garden productions ; more, however, I think, to cross breeding than to muling. It was entirely by the first of these processes that have been so greatly multiplied and improved our fruits for the dessert, and the gay flowers that adorn our gardens. The Pelargonium, the Calceolaria, the Dahlia, the Verbena, and a thousand others — what would they be, but simple wild flowers, without the power of man exercised in this way ? " To the cultivators of ornamental plants," says Mr. Herbert *, "the facility of raising hybrid varieties affords an endless source of interest and amusement. He sees in the several species of each genus that he possesses the materials with which he must work, and he considers in what manner he can blend them to the best advantage, looking to the several gifts in which each excels, whether of hardiness to endure our seasons, of brilliancy in its colours, of delicacy in its markings, of fragrance, or stature, or profusion of blossom ; and he may anticipate, with tolerable accuracy, the probable aspect of the intermediate plant which he is permitted to create : for that term may be figuratively applied to the introduction into the world of a natural form which has probably never before existed in it. In constitution the mixed offspring appears to partake of the habits of both parents ; that is to say, it will be less hardy than the one of its parents which bears the greatest exposure, and not so delicate as the other : but, if one of the parents is quite hardy, and the other not quite able to support our winters, the probability is, that the offspring will support them, though it may suffer from a very unusual depression of the thermometer, or excess of moisture, which would not destroy its hardier parent."

* See much the most valuable and practical account of cross breeding and muling which has been yet published in regard to horticulture, in the *Amaryllidaceæ* of the Hon. and Rev. W. Herbert, p. 335. et seq.

In the many successful attempts made by Mr.
Knight to improve the quality of fruit trees by
raising new varieties, his method was to obtain
crossbreds by fertilising the stigma of one variety
of known habits with the pollen of another also of
known habits.   But, in doing this, his experiments
were not conducted at random, and without due consideration; on the contrary, we learn from himself,
that he was very careful in selecting the parents
from which his crossbreds were obtained.   He found
that the general opinion, that the offspring of crossbred plants as well as crossbred animals usually
presents great irregularity of character, is unfounded;
and that if a male of permanent habits, and of
course not crossbred, be selected, that will completely overrule the disposition to sport, " the permanent character always controlling and prevailing
over the variable."   He tells us that he usually
propagated from the seeds of such varieties as are
sufficiently hardy to bear and ripen their fruit, even
in unfavourable seasons and situations, without the
protection of a wall ; because, in many experiments
made with a view to ascertaining the comparative
influence of the male and female on their offspring,
he had observed in fruits, with few exceptions, a
strong prevalence of the constitution and habits of
the female parent.   Unfortunately, however, this is
precisely the reverse of the result at which Mr.
Herbert has arrived in the very great number of

experiments performed by himself on that subject, he believing that the male parent generally influences the character of the foliage, and the female that of the flowers (*Amaryllidaceæ*, p. 348. 377.); and although it does appear to me that, in the majority of cases, Mr. Herbert's opinion is the more correct of the two, yet I fear there is too little certainty in the results of hybridising to justify the establishment of any axiom upon the subject.

This power of muling, properly so called, is confined within very narrow limits, and can hardly be said to exist at all between species of different genera, unless under that name are comprehended some of the spurious creations of inconsiderate botanists. There are, indeed, many cases of species very closely allied to each other which it is either impossible to mule, or so difficult that no one has yet succeeded in effecting it. Mr. Knight never could make the Morello breed with the common Cherry. I have in vain endeavoured to mule the Gooseberry and Currant, and we do not possess any garden production known to have been produced between the Apple and the Pear, or the Blackberry and the Raspberry, any of which might have been expected to intermix. As to mules obtained between plants of distinct genera, we have, no doubt, upon record, some experiments said to have been performed successfully in crossing a Thorn-Apple with Tobacco, the Pea with the Bean, the Cabbage with the Horseradish, and so on ; but Mr. Herbert regards these cases, and I think with great reason, as apocryphal, and not to be relied on ; the fact being, as he truly states, " that in this country, where the passion for horticulture is great, and the attempts to produce hybrid intermixtures have been very extensive during the last fifteen years, not one truly bigeneric mule has been seen."

On the other hand, cross breeding (89.) will take place quite as readily among plants as among animals, and it is difficult to estimate the alteration which this process has really produced, although unperceived by us, in the amelioration and alteration of long-cultivated plants. We cannot reasonably doubt that a process so simple as that of dusting the stigma of one plant with the pollen of another, which must be continually happening in our gardens, either through the agency of insects or the currents in the air, and which, where it takes place between two varieties allied to each other, must necessarily produce a cross, — we cannot suppose, I say, that this occurs in our crowded gardens and orchards at that time only when we perform it artificially.

The operation itself, although so simple, consisting in nothing more than applying the pollen of one plant to the stigma of another, nevertheless requires to be guarded by some precautions. In the first place, it is requisite that the flower whose stigma is to be fertilised should be deprived of its own anthers before they burst, otherwise the stigma will be self-impregnated, and although superfœtation is not, by any means, impossible, yet it is not very likely to occur. Then, again, the application of the stranger pollen should be made at the time when the stigma is covered with its natural mucus; if not, the pollen will not act, either in consequence of the necessary lubrification of itself being withheld, from the stigma being too young, or because the stigma, from age, has lost its power of receiving the action of the pollen. Neither should the stigma be in any way injured after fertilisation has apparently taken place. The art of fertilisation consists in the emission, by the pollen, of certain tubes of microscopical tenuity, which pass down the style, and eventually reach the young seed, with which they come in contact; and, unless this contact takes place, fertilisation misses. Now the transmission of the pollen tubes from the stigma to the ovule, through the solid style, is often very slow, sometimes occupying as much as a month or six weeks, as in the Mistletoe.

Those who occupy themselves in attempts at improving the quality of cultivated plants should be aware of this; namely, that the real quality of either the fruit or the flower of a seedling cannot be ascertained when they are first produced; for it is only as plants advance in age that the secretions necessary for the perfect production of either the one or the other are elaborated.   Of this fact, the first produce of the Black Eagle Cherry tree afforded a striking example.   A part of it was sent, with other cherries, to the Horticultural Society ; and it was then, in the Fruit Committee, pronounced good for nothing.   It was so bad, that Mr. Knight, who raised it, would most certainly have taken off the head of the tree and employed its stem as a stock, but that it had been called the property of one of his children, who sowed the seed which produced it, and who felt very anxious for its preservation.   It has now become one of the richest and finest fruits of its species which we possess.

It may be expected that some mention should here be made of double flowers, and of the manner in which they are to be obtained.   But I confess myself unable to discover, either in the writings of physiologists, or in the experience of gardeners, or in the nature of plants themselves, any sufficient clue to an explanation of the causes to which their origin may be ascribed.   There are, however, several facts apparently connected with the subject, which deserve mention.

A double flower, properly so called *, is one in

* What is called a Double Dahlia is misnamed ; and so are all so-called double Composite flowers.   The appearance of doubling is caused in these plants by a mere alteration of the florets of their disk into the form of florets of the ray ; a very different thing from double flowers. (83.)

which the natural production of stamens or pistils
is exchanged for petals, or in which the number
of the latter is augmented without any disturbance of the former; in other words, it is a case
of the loss, on the part of a plant, of the power
necessary to develope its leaves in the state of
sexual organs. (83, 84.) But what causes that loss
of power we do not know. It can hardly be a
want of sufficient food in the soil; for double
flowers (the Narcissus, for instance) become single in very poor soil. On the other hand, it can
scarcely be excessive vigour; for no one has ever
yet obtained a double flower by promoting the
health or energy of a species. When plants
are excessively stimulated by unusually warm
damp weather at the period of flowering, their
flowers in such cases sometimes become monstrous:
but the effect of this is to lengthen their axis of
growth, and to form true leaves instead of floral
organs (84. fig. 14.), just the reverse of what occurs in a truly double flower; the varieties of
Rosa gallica often exhibit this kind of change. In
damp cloudy summers, some flowers assume the
appearance of being double, by the change of their
sexual organs into small green leaves, as occurred
very generally to Potentilla nepalensis in the
summer of 1839, a representation of which is
given at page 62.; but there was, at the same
time, scarcely a trace of any tendency, on the part

of those leaves, to assume the colour or texture of petals.

There is, evidently, a greater tendency in some flowers to become double than in others, and especially in those having great numbers of stamens or pistils. All our favourite double flowers, Hepaticas, Pæonies, Camellias, Anemones, Roses, Cherries, Plums, Ranunculuses, belong to this class; and, in proportion as the natural number of stamens diminishes, so do both the disposition to become double, and the beauty of the flowers when altered. The Pink and Carnation with ten stamens are the handsomest race next to those just mentioned; while the Hyacinth, the Tulip, the Stock, and the Wallflower with six stamens, and the Auricula and Polyanthus with five, form altogether an inferior race, if symmetry of form, and regularity of arrangement in the parts of the flower, are regarded as beauties of the highest order. If the mere circumstance of a plant having but a small number of stamens be a bar to its beauty when made double, how much greater an obstacle to it must be the natural production of unsymmetrical flowers. This occurs in the Snapdragon, which, with a five-lobed corolla, has but four stamens; and the consequence is, that, when it becomes double, the flower is a confused crowd of crumpled petals issuing from the original corolla.

I have heard of attempts to produce double

flowers by artificial processes, but I never heard of the smallest success attending such cases, unless the tendency to their production had already manifested itself naturally; as in the Stock, which will frequently become single from having been double, in which case its original double character may be recovered. A mode of effecting this has been described by Mr. James Munro. (*Gard. Mag.*, xiv. 121.) Having a number of Single Scarlet Ten-week Stocks, he deprived them of all their flowers as soon as he found that five or six seed-vessels were formed upon each spike, by which means he compelled all the nutritive matter that would have been expended upon the whole flower-spike and its numerous seed-vessels to be concentrated in the small number which he left; and the result, he says, was, that from the seed thus saved he had more than 400 Double Stocks in one small bed.

There can, I think, be no doubt that, if any original change to a double flower can possibly be effected by art, it will be more likely to occur with respect to those species which have an indefinite number of stamens, where the tendency to this monstrosity already exists. It is not many years since the Chryseis (Eschscholtzia) californica, a polyandrous plant, was introduced to our gardens; and I, at one time, made some attempts to render it double, conceiving it a good subject for experiment on that account, but I had no success; it has, however, accidentally become semi-double in Mrs. Marryat's garden, at Wimbledon ; and I entertain no doubt that seed skilfully saved from that plant would present its flowers in a still more double condition.

CHAP. XIX.

OF RESTING.

A GARDENER is said to rest a plant when he exposes it to a condition in which it cannot grow, and which is analogous to its winter state. For many parts of gardening, especially what relates to forcing and the management of exotic plants, this is a subject of the first importance.

If we look over the different climates of the world, we shall find that in each there are a season of growth, and a season in which vegetation is more or less suspended ; and that these periodically alternate, with the same regularity as our summer and winter. I do not know that there is in nature any exception to this rule : for even in the Tierra templada of Mexico, where it is said that, at the height of 4000 to 5000 feet, there constantly reigns the genial climate of spring, which

z

does not vary more than 8° or 9°, intense heat and excessive cold being alike unknown, and the mean temperature varying from 68° to 70°, we cannot suppose that, even in that favoured region, a season of repose is wanting; for it is difficult to conceive how plants can exist, any more than animals, in a state of incessant excitement. Indeed, it is pretty evident that these countries have a period when vegetation ceases; for Xalapa belongs to the Tierra templada, and we know that the Ipomœa purga, an inhabitant of its woods, dies down annually like our own Convolvuli.

But, although all plants have naturally a season of repose, their winter is not in all cases cold. In the tropics it is marked by coolness and dryness, while the summer is rainy and very hot; and in extra-tropical countries the two seasons vary in their character, according to latitude and local circumstances.

In some parts of Persia, Armenia, and Mesopotamia, the summer heats are excessive, while the winters are rendered cold by the proximity of mountains. Bagdad is described as having a cold winter, because of the proximity of the mountains of Koordistan; yet its heats are intense : in August, 1819, the thermometer stood at 120° in the coldest parts of the house, and at 108° at midnight in the open air. This was preceded by heavy rains, which raised the Euphrates $7\frac{1}{2}$ feet above the ordinary level: the whole country was like a vapour bath, and multitudes of persons dropped down dead: twenty-two in three days in a single caravan. In the northern provinces of Mexico the winters are of German rigour, while the summers are those of Naples and Sicily; the Tierra fria of that country has however a very different climate, the mean heat of the summer being 76°, and the winters so mild that the thermometer only occasionally falls below 32°.

At the Cape of Good Hope there are districts in which the period of wet is long and very severe; and many of the favourite flowers of our gardens are produced by those districts. The Karroos are plains of great extent, destitute of running water, with a soil of clay and sand, coloured like yellow ochre by the presence of iron, and lying on the solid rock. During the dry season the rays of the sun reduce the soil nearly to the hardness of brick: Fig Marigolds, Stapelias, and other fleshy plants, alone remain green; nevertheless, the bulbs and tribes of Iridaceous and other plants are able to survive beneath the sun-scorched crust, which appears indeed to be necessary to their nature. But in the wet season these bulbs are gradually reached by the rain; they swell beneath the earth; and at last develope themselves so simultaneously that the arid plains become at once the seat of a charming verdure. Presently afterwards, myriads

of the gay flowers of the Iridaceæ and Mesem-
bryanthemums display their brilliant colours : but
in a few weeks the verdure fades, the flowers
disappear, hard dry stalks alone remain ; the hot
sun of August, when in those latitudes the days
begin to lengthen, completes the destruction
of the few stragglers that are left, the Karroo
again sinks into aridity and desolation, and the
desert reappears.   What succulents survive are
covered with a grey crust, and derive their nourishment only from the air.   In other parts of the
Cape of Good Hope the mean range of the thermometer in winter is 48° to 93°, with cold rain, while
that of the summer is from 55° to 96°, with dry
days and damp nights.

In the Canaries we have the season of growth
from November to March, when rains fall like
those of Europe, and the mean temperature is
66°; and the period of rest is April to October,
when it never rains, and the mean temperature is
73°.

In Brazil the seasons are thus described by Mr.
Caldcleugh : — " The summer begins about the
months of October or November, and lasts until
March or April.   This is the wet season ; but the
rains by no means descend from morning till night,
as in some other tropical countries, but commence,
generally, every afternoon about four or five o'clock
with a thunderstorm.   The heaviness of the rain

can only be conceived by those who have been in these latitudes. This fall naturally arrests the sea breeze, and the succeeding night is dark and cloudy. Formerly these diurnal rains came on with such regularity that it was usual, in forming parties of pleasure, to arrange whether they should take place before or after the storm. During this period of the year there is seldom, if ever, a deposition of dew. From April until September very little rain falls; vegetation almost stops, and, to the eye of every one who has not just arrived from Europe, a wintry appearance is discernible. The land and sea breezes do not succeed each other with the same regularity, and are, besides, more frequently disturbed by violent gusts from the s. w., imagined to be the tails of those destructive winds the Pamperos of the River Plate. The nights are beautifully clear; Venus casts a shadow, and the southern constellations are seen in all their beauty. The dews, as might be expected, are at this season very copious." (*Brande's Journal*, No. 27. p. 41.)

In other parts of the tropics the seasons of growth and rest are equally marked. In Ava, during the rainy season, which lasts from May to October, the mean temperature varies from $78°$ to $91·5°$; while, in the dry season, from November to April, it falls to from $63°$ to $80°$. At Calcutta, in the growing season, from April to October, 58 inches of rain commonly fall, with a mean temperature of 79° to 86°; while during the season of wet, from November to March, there is not perhaps above an inch of rain, and the thermometer sinks to from 66° to 80°. At this time vegetation is said, in such countries, to "labour under a deadly languor; but one night's rain converts an arid plain into a verdant meadow."

In most of the West India Islands situated under the tropic of Cancer, there is said not to be much difference in the climate, so that accurate observations made on any one of them may be applied with little variation to them all. Malte Brun gives the following sketch of their seasons. "The spring begins about the month of May; the savannas then change their russet hue, and the trees are adorned with a verdant foliage. The periodical rains from the south may at this time be expected; they fall generally about noon, and occasion a rapid and luxuriant vegetation. The thermometer varies considerably; it falls sometimes six or eight degrees after the diurnal rains; but its medium height may be stated at 78° Fahrenheit. After these showers have continued for a short period, the tropical summer appears in all its splendour. Clouds are seldom seen in the sky; the heat of the sun is only rendered supportable by the sea breeze, which blows regularly from the south-east during the greater part of the day. The nights are calm and serene, the moon shines more brightly than in

Europe, and emits a light that enables man to read the smallest print; its absence is in some degree compensated by the planets, and, above all, by the luminous effulgence of the galaxy. From the middle of August to the end of September, the thermometer rises frequently above 90°, the refreshing sea breeze is then interrupted, and frequent calms announce the approach of the great periodical rains. Fiery clouds are seen in the atmosphere, and the mountains appear less distant to the spectator than at other seasons of the year. The rain falls in torrents about the beginning of October, the rivers overflow their banks, and a great portion of the low grounds is submerged. The rain that fell in Barbadoes in the year 1754 is said to have exceeded 87 inches. The moisture of the atmosphere is so great, that iron and other metals easily oxidated are covered with rust. This humidity continues under a burning sun; the inhabitants, (say some writers) live in a vapour bath." (*Malte Brun's Geography*, vol. v. p.569., Eng. ed.)

It is evident, from what has been said, that the natural resting of plants from growth is a most important phenomenon, of universal occurrence, and that it takes place equally in the hottest and the coldest regions. It is, therefore, a condition necessary to the wellbeing of a plant, not to be overlooked under any circumstances whatever; and there cannot be any really good gardening where this is not attended to in the management of plants under glass. Rest is effected in one of two ways; either by a very considerable lowering of temperature, or by a degree of dryness under which vegetation cannot be sustained.

The way in which the physical powers of vegetation are affected by this has been already explained (114.); and, in practice, it is found a point of the utmost consequence. The early fruit-gardener draws his Vines out of the vinery, and takes the sashes from his Peach and other forcing-houses, when the artificial season of growth is over, in order to prepare them for the duty of a succeeding season; although this operation is performed in summer, its effect is to expose them to dryness, which arrests their growth, and favours the deposit in their wood of the matter required for the produce of a succeeding year.

The effects of a very dry atmosphere are necessarily an inspissated state of the sap of the plant; and this in all cases leads to the formation of blossom-buds and of fruit. It thus operated upon some Pine-apple plants in Mr. Knight's garden, to such an extent as to cause even the suckers from their roots to rise from the soil with an embryo pine-apple upon the head of each, and every plant to show fruit, in a very short time, whatever were its state and age. Very low temperature, under the influence of much light, by retarding and diminishing the expenditure of sap in the growth of plants, comparatively with its creation, produces nearly similar effects, and causes an early appearance of fruit.

The operations of forcing are essentially influenced by these facts; and, by a skilful alteration of the periods of rest, we are enabled to break in upon the natural habits of plants, and to invert them so completely, that the flowers and fruits of summer are obtained to load our tables even in winter. Of this, the following instance, taken from a paper by Mr. Knight in the *Horticultural Transactions* (vi. 232.), is a sufficient illustration.

" A Verdelho Vine, growing in a pot, was placed in the stove early in the spring of 1823, where its wood became perfectly mature in August. It was then taken from the stove and placed under a north wall, where it remained till the end of November, when it was replaced in the stove; and it ripened its fruit early in the following spring. In May it was again transferred to a north wall, where it remained in a quiescent state till the end of August. It then vegetated strongly, and showed abundant blossom, which, upon being transferred to the stove, set very freely; and the fruit, having been subjected to the influence of very high temperature, ripened early in the month of February."

The strawberries of February and March are in like manner procured by exposing the plants to such an amount of dryness and heat as can be obtained by presenting them unwatered, in pots, to the sun, at an early period of summer; so as to cause a sufficient accumulation of excitability by the end of autumn, instead of the month of May.

It must be manifest that the operations of the flower-gardener should be regulated by the same principles, although it must be confessed that they are often little considered; a circumstance the more strange, from the indispensable necessity of resting fruit trees being universally known. It is to the giving their plants the proper kind of rest that some gardeners owe the magnificent blossoming of their Chinese Azaleas, Cacti, Camellias, and other forced flowers, much more than to any peculiarity in the compost they employ, which is often a point of subordinate interest, although generally regarded as of the first importance. If but little progress has as yet been made by art in altering the time of flowering of particular races, so as to invert their seasons, this is certainly very far from being beyond the reach of attainment; and there is apparently no more reason why a Chinese Chrysanthemum should not be compelled to flower at midsummer instead of November, or a Dahlia at Christmas, than that Vines and Strawberries should ripen fruit in February. The great difficulty to contend against in obtaining winter flowers is want of light; but, by the employment of slender iron sash-bars and large glass, a sufficient amount of this important vital agent may be obtained in England even at that season of the year.

But it is not merely the periodical rest of winter and summer that plants require; they have also their diurnal repose : night and its accompanying refreshment are as necessary to them as to animals. In all nature the temperature of night falls below that of day, and thus one cause of vital excitement is diminished; perspiration is stopped, and the plant parts with none of its aqueous particles, although it continues to imbibe them by all its green surface as well as by its roots ; the processes of assimilation are suspended ; no digestion of food and conversion of it into organised matter takes place ; and, instead of decomposing carbonic acid by the extrication of oxygen, they part with carbonic acid, and rob the air of its oxygen, thus deteriorating the air at night, although not to the same amount as they purify it during the day. It is, therefore, most important, that the temperature of glass houses should, under all circumstances whatever, be lower than that of the day ; and it is probable that this ought to take place to a greater extent than is generally imagined by even the best practical gardeners. We are told that, in Jamaica and other mountainous islands of the West Indies, the air upon the mountains becomes, soon after sunset, chilled and condensed, and, in consequence

of its superior gravity, descends and displaces the
warm air of the valleys ; yet the sugar-canes are so
far from being injured by this decrease of temperature, that the sugars of Jamaica take a higher
price in the market than those of the less elevated
islands, of which the temperature of the day and
night is subject to much less alteration. At Fat-
tehpúr, in the East Indies, the difference in temperature between night and day amounts to as
much as 78°, on an average of the whole year; in
April the greatest heat by day is 110°, that of
night is only 65° ; in January the thermometer
falls to 38° at night, while the day is 76°; and
there are 40 degrees of difference between the day
and night in May, one of the hottest months, when
the thermometer ranges as high as 115°. At
Calcutta, in May, the thermometer averages 93°
in the day, and 79° at sunrise; while in January the
temperatures are 77° and 56° respectively, for those
two periods.

When we compare these facts with the habits
of plants just adverted to, we must, I think, see
that it is the purpose of nature to reduce the force
which operates upon the excitability of vegetation
at that period of the twenty-four hours, when, from
other causes, the powers of digestion and assimilation are suspended. As far as is at present known,
that power is heat ; and therefore we must suppose
that, to maintain at night in our hot-houses a tem-

perature at all equal to that of the day, is a practice to be much condemned. Plants will no doubt lengthen very fast at night in a damp heat, but what is at this time produced seems to be a mere extension of the tissue formed during the day, and not the addition of any new part; the spaces between the leaves are increased, and the plant becomes what is technically and very correctly called drawn; for, as has been justly observed, " the same quantity only of material is extended to a greater length, as in the elongation of a wire."

Mr. Knight has pointed out another ill effect of high temperature during the night, namely, that it exhausts the excitability of a tree much more rapidly than it promotes its growth, or accelerates the maturity of its fruit; which is, in consequence, ill supplied with nutriment at the period of its ripening, when most nutriment is probably wanted. The muscat of Alexandria, and other late grapes, are, owing as he thinks to this cause, often seen to wither upon the branch in a very imperfect state of maturity; and the want of richness and flavour in other forced fruits is often attributable to the same cause. " There are few peach-houses," he adds, " or indeed forcing-houses of any kind, in this country, in which the temperature does not exceed, during the night, in the months of April and May, very greatly that of the warmest valley in Jamaica in the hottest period of the year. There are pro-

bably, as few forcing-houses in which the trees are
not more strongly stimulated by the close and damp
air of the night, than by the temperature of the
dry air of the noon of the following day.   The
practice which occasions this cannot be right : it
is in direct opposition to nature."   In the same
paper from which the foregoing is an extract
(*Hort. Trans.*, ii. 135.), the same great experi-
mentalist records the result of his own management
of a peach-house, where a due regard was had to
the preservation of a sufficiently low temperature
at night.   " As early in the spring as I wanted the
blossoms of my Peach trees to unfold, my house
was made warm during the middle of the day ; but
towards night it was suffered to cool, and the trees
were then sprinkled, by means of a large syringe,
with clear water, as nearly at the temperature at
which that usually rises from the ground, as I could
obtain it ; and little or no artificial heat, was given
during the night, unless there appeared a prospect
of frost. Under this mode of treatment, the blossoms
advanced with very great vigour, and as rapidly as
I wished them, and presented, when expanded, a
larger size than I had ever before seen of the same
varieties ; which circumstance is not unimportant,
because the size of the blossom, in any given
variety, regulates, to a very considerable extent,
the bulk of the future fruit."

CHAP. XX.

OF SOIL AND MANURE.

NOTWITHSTANDING all that has been written upon these substances, and the endless accounts we possess of their real or supposed action upon vegetation, I must confess that the contradictions are so numerous, the exceptions to supposed rules so frequent, and physiology is so insufficient to account for the greater number of well ascertained facts, that it does not appear to me possible to construct any tolerable theory relating to them.

Mr. Knight has observed that varieties of the same species of fruit tree do not succeed equally in the same soil, or with the same manure : the Peach in many soils acquires a high degree of perfection, where its variety, the Nectarine, is of comparatively little value ; and the Nectarine frequently possesses its full flavour in a soil which does not well suit the Peach. The same remark is also applicable to the Pear and the Apple ; and, as defects of opposite kinds occur in the varieties of every species of fruit, those qualities in the soil which are beneficial in some cases will be found injurious in others. In those districts where the Apple and Pear are cultivated for cider and perry, much of the success of the planter is found to depend on his skill or good

fortune in adapting his fruits to the soil. (*Hort.*
Trans., i. 6.) Rhododendrons and Kalmias are
usually cultivated in peat earth mixed with sand,
and yet they grow as well in fresh hazelly loam,
without any mixture whatever; and, than these
two kinds of soil, none can be apparently more dis-
similar. The fine American cottons are grown in a
calcareous sand, those of India in a deep black
saponaceous earth : the American cotton will not
thrive in the latter, nor that of India in the former,
as has now been ascertained; and yet the species of
Gossypium producing the two qualities have no
organic differences which can, so far as has yet been
ascertained, explain in the smallest degree the
necessity, under which it is evident that they labour,
of being provided with different kinds of food.
The Alnus glutinosa, or Common Alder, flourishes
in wet clayey meadows ; while Alnus incana, or
Upland Alder, is equally suited to a dry and light
land : we are totally ignorant of the reason of such
a case as this. Rhododendron hirsutum and Erica
carnea are, in their wild state, confined to calcareous
soil; while Rhododendron ferrugineum grows exclusively on granite, and Erica vagans on serpentine.
We are informed by Beyrich (*Gardener's Magazine*,
iii. 442.) that " the Pine-apple, in its wild state, is
found near the sea-shore ; the sand accumulated
there in downs serving for its growth, as well as for
that of most of the species of the same family.

The place where the best Pine-apples are cultivated
is of a similar nature. In the sandy plains of Praya
velha and Praya grande, formed by the receding of
the sea, and in which no other plant will thrive,
are the spots where the Pine-apple grows best.
The cause of this lies evidently in the composition
of the sand, which chiefly consists of salt, lime from
decomposed shells, and a very little vegetable
mould. Warmth, lime, salt, and moisture, seem
therefore to be the principal ingredients in which
the Pine-apple thrives. Sand will take a very high
and continued degree of warmth, being often heated
by the sun so much as to scorch vegetation, and
yet it seldom dries to a greater depth than from
eight inches to one foot. Sea salt is well known
for its property of attracting the nocturnal damps,
and retaining them a long time. The lime of the
shells seems to be the principal manure, which has
also been proved by the English here, who, by
manuring their Pine-apples with a mixture of
stamped oyster-shells and vegetable earth, produce
very large fruit. The natural mould, usually
slightly mixed with sand, is partly of a vegetable,
and partly of a mineral, origin." But it is well
known that the Pine-apples of England are much
superior to those of South America, and yet English
gardeners grow their plants neither in sand, nor
saline nor calcareous soil. As to manures, some
plants bear them in almost any quantity, others

suffer from the access of only a small quantity.
The Vine and the Mulberry can hardly be over-
manured, no soil was ever found too rich for Roses :
but Coniferous plants can scarcely bear any manure, and the Peach is often greatly injured by excess of it in a solid state ; yet this same plant will
bear a very considerable quantity in a liquid form.

The application of soils and manures to plants
must, therefore, remain at present exclusively
within the domain of *art*. There are, however,
some general remarks which it is possible to offer
with tolerable confidence.

Soil, considered without reference to the organisable substances it contains, appears to act upon
plants chiefly by its power of absorbing and parting
with heat and moisture. When soil is tenacious,
or plastic, it absorbs heat slowly, and it parts with
its water with great difficulty, as is the case in the
London clay ; the number of cultivated plants to
which this is suitable is so small that it is almost
expelled from gardens, where the object is to
expose the cultivated species to conditions more
favourable than those afforded them by nature. The
small amount of bottom heat afforded by clay, and
the impossibility of effectually draining it, sufficiently
explain the badness of its quality for gardening
purposes, even without taking into account the
difficulty experienced by plants in rooting in it,
from the resistance afforded to the passage of the

spongioles by so compact a substance. On the other hand, loose sand, whose particles have no cohesion, although it imbibes water with great facility, parts with it as readily, and, being easily heated by the sun's rays, becomes so soon dried up as to be for that reason as unsuitable to most plants as plastic clay itself. It is by obtaining a mean between these two extreme cases that the soil is formed most favourable to the growth of plants in general; hence the mixtures of peat, loam, and sand, which are so continually employed. These substances counteract each other's influences, the loam by consolidating the sand, and the sand by lightening the loam, and the peat by binding them all together, and preserving their perfect admixture, independently of its manuring qualities. It is, however, a well ascertained fact, that loam containing a considerable quantity of calcareous matter is in general much better suited to cultivation than such as is destitute of it : the reason for which seems to be, in part, that calcareous earth enters largely into the organisation of all plants, in which it is deposited in the state of the oxalate and phosphate of lime ; and, in part, because, as was shown by Davy, there is a strong action between the lime and vegetable matter contained in soils, the result of which is a compost partly soluble in water.

Doubtless one of the safest rules for a gardener, in determining the soil required for a given plant, would be, if practicable, to ascertain what amount of mineral matters it contains, and to select earth in which those substances abound. For, although it may be asserted that the presence of iron, copper, or other substances, in plants, in minute quantities, is accidental and unimportant, yet such a supposition is gratuitous, if not altogether unfounded; for I do not know what warrant we have for saying that any of the constant phenomena of nature, however minute they may seem to be, are accidental. This at least is certain, that, where mineral substances occur abundantly in plants, they are part and parcel of their nature, just as much as iron and phosphate of lime are of our own bodies; and we must no more suppose that grasses can dispense with silica in their food, or marine plants with common salt, than that we ourselves could dispense with vegetable and animal food. Flint is found on the exterior of the whole Graminaceous order, without exception; it forms the polished surface of the Cane Palm, the grittiness of many kinds of timber; sulphur abounds in Cruciferous plants, especially Mustard; copper in Coffee *, Wheat, and many other plants (it is believed in the state of a phosphate); iron, as a peroxide,

---

* Seventy millions of kilogrammes of coffee arrive annually in Europe; of these, 560 kilogrammes consist of copper, according to M. Sarzeau. The weight of copper consumed in bread in France is 3650 kilogrammes annually. (*De Candolle, Physiologie Végétale*, p. 389.)

in Tobacco. John, in his experiments upon these matters, found that the Ramalina fraxinea and Borrera ciliaris, two lichens, contained a great quantity of the last metal, although he could not find a trace of it in the Fir tree, on the topmost branches of which the lichens grew. We cannot suppose that such things are the result of accident, and that it is unimportant to the plants containing minerals thus constantly, whether such substances are present in their soil or not.

Manures act apparently in one of three ways, either by merely stimulating the vital forces, as common salt; or by their power of absorbing moisture from the atmosphere, as salt and the muriate of lime, obtained by mixing together equal parts of salt and lime ; or by supplying the plant with soluble carbon and nitrogen. It is in proportion to their power of furnishing these principles, and to the length of time during which they continue to do so, that manures are active or sluggish, and durable or ephemeral in their operation. Carbonic acid, when decomposed, furnishes an essential part of the starch and other substances secreted by plants; and nitrogen seems, from its great abundance in their system, at least when young, to be indispensable to their existence : the first is a fact of universal notoriety, the latter has been ascertained by modern chemists to be also apparently a constant phenomenon. (See *Introduction to Bo-*

tany, 3d ed. p. 370. 379, &c., and Appendix.) For these reasons every description of putrefying animal or vegetable matter, from putrid yeast and malt dust to horses' hoofs and feathers, have been used for the purpose of fertilising land, the nature of whose different actions constitutes a study of itself, very obscure, but of the highest degree of importance.*

In the more delicate of horticultural operations, liquid manure, prepared by steeping dung in water, and drawing it off when clear and of the colour of porter, is most generally now employed, and is undoubtedly the best form in which it can be administered, in consequence of its concentration, the facility of its administration in any quantity, and its containing nothing but soluble matter. It was first used by Mr. Knight, who not only applied it with great advantage to Fruit trees, but also to Heaths

---

* DeCandolle gives, from Hermstädt's observations (*Annalen der Landwirthschaft*, xxii. 1.), the following results obtained by the action of different manures upon Rye, Barley, and Oats, under exactly equal circumstances : —

	Rye.	Barley.	Oats.
Sheep's dung	13-fold.	16-fold.	14-fold.
Goats' dung	13	15	15
Horse-dung	11	13	14
Cow-dung	9	11	16
Human fæces	13	13	14½
Pigeons' dung	9	10	12
Human urine	13	13	13
Dry bullock's blood	14	16	12½
Vegetable earth	6	7	13
Unmanured soil	4	4	5

and other flowers ; and it is, with the exception of bone dust, the form of manure best adapted to all plants in pots.

What most concerns the subject of this work is, not the nature of manure, but the proper time and manner of applying it to garden plants. Provided manure is of a permanent character, it does not very much matter at what time it is administered, because, if it does not act at first, it will sooner or later ; but when it is of such a nature as to be easily dissipated, like malt-dust, or soot, or yeast, a knowledge of the proper season becomes extremely necessary. Plants will not receive the influence of manure so readily at any season as when they are in the most rapid and steady growth ; because at that time the absorbing force of their roots, and their vital energies, are all greatest. It is for this reason that a top dressing is almost useless to a lawn at midsummer, but better in the spring, and best of all in October. If applied at midsummer, the ground is dry, the herbage closely shorn, and the vegetation extremely languid, partly in consequence of the constant operation of the mower, and partly because our summertide is the winter of herbage grasses, which only flourish in the cool and damp seasons of the year. When a top-dressing is applied in the spring, the lawn profits by it so long as it continues to grow vigorously; but the quick approach of summer daily interferes with the force of this kind of vegetation, and diminishes the effects of the manure. On the contrary, if October is the season chosen for the operation, the grasses are then beginning to grow steadily, the operations of the mower are, or should be, suspended, and there are seven clear months at least during which the effects of the manure continue to be felt.

It may be indifferent at what season such manures as bones, and other kinds of matter which decompose very slowly, are employed; yet there can be no doubt that upon every known principle they also should be given at a time when vegetation is most active; hence the every-day practice of digging manure into the borders of a garden in spring, or shortly before an annual crop is about to be committed to the soil.

As to the manner of applying manure, it must be obvious that it can be of no use unless it is in contact with the absorbing parts of the roots; now those parts are the young fibres and spongioles, as has been already stated (23, 24.), and, when plants have arrived at any considerable size, the roots form the radii of a circle whose circumference is the principal line of absorption. This being so, if a plant has arrived at the state of a bush or tree, it is useless to apply manure to the base of the stem, because that is precisely where the power of absorption is the weakest, if it exists at all; and, as the circle * formed by the roots is generally greater than that of the branches, the proper manner of applying manure is, to introduce it into the ground at a distance from the stem about equal to the radius formed by the branches. And yet, although this is so evidently right, I have seen a gardener, who ought to have known much better, sedulously administering liquid manure, by pouring it into the soil at the base of the stem ; which is much the same thing as if an attempt were made to feed a man through the soles of his feet.

* In Book I. par. 33. this is obscurely expressed, but corrected in the errata.

INDEX.

A.

ABSORPTION, force of, in spongelets, 17.
Acetate of lime, solution of, rejected by roots, 19.
Air, deterioration of, by plants at night, 347.
    introduction of heated, to plant compartments, 158.
    purification of, by plants during the day, 347.
    of plant-houses, ventilation a means of drying, 158.
Air passages penetrate plants in all directions, 105.
Alburnum, 29.
    its connexion with the spongelets, 12.
    offices performed by it, 33.
    its importance, 34.
    its quantity proportionate to the number of buds, 194.
Alburnous substance, 194.
Ammonia employed to promote germination, 171.
Amylaceous substance, 195.
Annuals, oleraceous, 120.
Annual rest of plants, 337.
Anther, 56.
Aquatic plants, 118.
    necessity of maintaining a due degree of temperature in the
      water in which they are grown, 113.
Aqueous matter, necessity of its excess being decomposed in fruits
    during the process of ripening, 120.
Aqueous particles in plants, effects of their being frozen, 89.
Arid regions, 339.
Atmosphere, temperature of, at various places, 94.
    unfavourable state of, a cause of sterility in flowers, 177.
    dry, produces an inspissated state of the sap, 344.
Atmospheric dryness, or moisture, extremes of, 134, 135.
                       averages of, 135.
    moisture, 128.

374 INDEX.

C C

W.

THE END.

Lᴏɴᴅᴏɴ :
Printed by A. Sᴘᴏᴛᴛɪsᴡᴏᴏᴅᴇ,
New-Street-Square.